LESSONS UNDER THE OAK TREE

LESSONS UNDER THE OAK TREE

Karen Kluger

Xlibris

Xlibris, Philadelpia

Handwritten inscription:

> Dear Paul,
> How precious you are to me, my only godson. Though we are miles apart, know you are always in my heart & prayers. Love, Karen "Joxi" Kluger

Copyright © 2005 by Karen Kluger.

Library of Congress Number: 2004195692
ISBN :   Hardcover   1-4134-5546-8
              Softcover   1-4134-5545-X

1. Nature—Inspirational   2. Ecopsychology
              3. Spirituality
4. Trees—Oak—California   5. Trees—Oregon
158—dc20

Photography by Russ Heyman

All rights reserved. No part of this book may be reproduced or transmitted in any form or by any means, electronic or mechanical, including photocopying, recording, or by any information storage and retrieval system, without permission in writing from the copyright owner.

Queries regarding rights and permissions should be addressed to:
karenkluger.com

This book was printed in the United States of America.
To order additional copies of this book, contact:
Xlibris Corporation
1-888-795-4274
www.Xlibris.com
Orders@Xlibris.com

For my sons
with love and gratitude for their
unending encouragement.

"I wish all to know that I do not propose to sell any part of my country, nor will I have the whites cutting our timber along the river. I am particularly fond of the little groves of oak trees. I love to look at them, because they endure the wintry storm and the summer's heat, and not unlike ourselves seem to flourish by them."

Sitting Bull,
Hunkpapa Sioux Warrior Chief
and Spiritual Leader

Contents

	List of Illustrations	xi
	Acknowledgements	xiii
	Introduction	1
1.	The Phone Call	3
2.	Discovering the Oak Tree	6
3.	The Standing Ones	10
4.	Wild Things	13
5.	Life in a Nutshell	16
6.	Soul Mates	20
7.	The Bonsai	25
8.	The Four Directions	28
9.	Wind Storm	31
10.	The Skater's Attitude	35
11.	Leaves of Red and Green	38
12.	Karmic Relationships	42
13.	Choices	46
14.	Celebrate	50
15.	Captain Jake	54
16.	Teacher of Love	59
17.	Nature's Art	63
18.	The Amazing Seed	70
19.	Creativity	77
20.	Living in the Moment	80
21.	The Ficus Vine	83
22.	The Espalier	86
23.	Sing a Happy Song	89

24.	Call the Doctor	94
25.	The Ring of Care	96
26.	New York City	100
27.	Woodpecker's Band	105
28.	Fish Food	107
29.	Leaving Something Behind	110
30.	Giving Back	113
31.	A Matter of Opinion	116
32.	Ignorance	120
33.	The Ancient Oak	122
34.	The Seed Reproduces Itself	126
35.	Squirrel	130
36.	An Earthy Woman	133
37.	Without One, No Other	139
38.	Eggs in the Nest	143
39.	The Forest of the Future	145
40.	A Gift to Give	149
41.	Toxin Gobblers	155
42.	Scales of Balance	159
43.	A Nice Neighborhood	165
44.	Nature Shows the Way	171
45.	The Drought	179
46.	All Things Connect	186
47.	The Emerald Forest	190
48.	The Lover's Forest	194
	Selected Bibliography	199

List of Illustrations

Cover: Coast live oak, Malibu Canyon, CA

1. Mediterranean cork oak, Westlake Village, CA 7
2. Coast live oaks on hillside, Malibu Canyon, CA 15
3. Douglas fir tree, Lake Oswego, OR 41
4. Valley oak, Westlake Village, CA 47
5. Young oak tree, Thousand Oaks, CA 53
6. Oaks along Malibu Creek, Malibu Canyon, CA 67
7. Oak along Route 128 near Cloverdale, CA 98
8. Panorama of oak covered foothills, Thousand Oaks, CA 127
9. Coast live oak above dam, Malibu Canyon, CA 160
10. Oak along Highway 101 near Geyserville, CA 162
11. Coast live oaks in Malibu Canyon, CA 166
12. Coast live oaks in Malibu Canyon 169
13. Nursing log, Tyron Creek State Park, Dunthorpe, OR 176
14. Oaks along Highway 101 near Geyserville, CA 184
15. Valley oak, Malibu Canyon, CA 196

Acknowledgements

Grateful acknowledgement is made to the following people.

A special thanks to those at Xlibris: Jim Grammond, Toni Cabatingan, and cover designer Thomas McAteer.

Thank you also to members of my critique group for their valuable advice: Jennifer Bernier, Dade Cariaga, the late Jack Myers, Eric Ottem and Cheryl Long Riffle. To fellow writer and horticulturist Herbert Piekow a special thanks for his kindness and support. Much appreciation is given to my dear friend Paul for his humor, affection and sage advice. Appreciation goes to Teri Jochums and Greg Colliton for their interest and encouragement.

A special note of gratitude to close friends: Jennifer Bernier, Cathy O'Connor, Jan McKelvey, MaryLou Kenworthy, and Lois Bernstein, each of whom listened to the ideas as they came and provided a wise sounding board, and Marjorie Rosen, who started me writing.

A heartfelt thanks goes to Russ Heyman for his love and encouragement, as well as the photographs in this book. And, last but never least, unending gratitude goes to my agent and editor Randy Heyman for his technical expertise, remarkable insight and suggestions as he guided the manuscript toward publication.

Introduction

To the reader from the author,

While walking near my home, I was drawn to an impressive old oak tree. On my walks, I frequently stopped by the tree. In the simple act of leaning against the oak with my arms resting on its lowest branches—as if it were a fence and I was talking to a neighbor—life began to take on new meaning. The insights came at a time when I had to stop and see everything around me with new eyes, including the structure upon which I'd built my life.

Nature became my teacher, guiding my heart and mind toward a more spiritual place. I found in nature what the philosopher Johann Wolfgang von Goethe called "the living, visible garment of God." I don't believe my experience is unique. Sometime during every life, nature draws us near then through familiarity we develop a fondness for, say, a river, a garden, an animal, a mountain, or the stars in the night sky. Then if we ask, nature takes our hand. When we open our mind and listen with our heart, nature will share her secrets with us. I suspect everything God thought would be helpful for us to know in order to live and find happiness is written in nature.

As time passed, I began keeping a journal about new insights that came while walking among the oaks. The collection of insights eventually became this book, which uses trees to illustrate nature's wisdom and spirituality.

Some stories are about forests, which serve as a metaphor for society. Trees of various shapes and sizes fill the forest. The tree's branches cover a large area, which allows their seeds to be broadcast far and wide. The seeds are like ideas that grow strong, becoming deeply rooted in a culture. Some ideas live a long time, sometimes for centuries, and affect society's judgment about how to care for the land and all that live upon it.

A large percentage of the narratives cite specific trees, which serve as a metaphor for you and me. These stories contain insights into the wisdom, love, and liberty that lays hidden in nature. By learning to read nature, we find depths to which nature honors truth and life itself. To every season and everything created there is beauty. That splendor exists in a rose, a snowflake, the view from a mountaintop, in the sea, and in colorful bright autumn leaves. Just as surely, that beauty resides in each of us. Sometimes the grandeur may appear to be hiding, but it is always there ready to come forth.

Although some names in the story have been changed, the stories are based on true incidents that highlight the deep connections between nature and human nature. Lastly, *Lessons Under the Oak Tree* tells the story of how I learned that most of what I need to know for a happy, meaningful life is as simple as a tree, and not as complicated as the forest.

—*Karen Kluger*

Chapter 1

THE PHONE CALL

The loud ring of the telephone woke me from a deep sleep, yet it was that phone call that led me on a journey of discovery that has not yet ended.

The call was from Nick, a talented man respected by many for what he is inside. The outside is rather appealing also with his blond hair, blue eyes, and ready smile. He's the kind of guy that has never had trouble getting women. With his deep mellow voice, he said, "A couple weeks ago, I met an *absolutely lovely woman*."

I was instantly awake and plumped the pillows to sit up in bed.

"I met her at a party. She's due here any minute, so I can't talk long. We're going to San Francisco."

"Sounds like fun."

His voice began to resonate with passion as he described the relationship. "It's *so karmic*. I feel a strong sense of destiny."

Remembering that feeling, I replied, "It happens. Sometimes you meet someone and you feel you know that person. And, they have the same impression about you."

"That's exactly how this is." As the conversation continued, it was obvious they enjoyed each other totally in public and private recreation. I surmised all six feet of him of no longer walked, but drifted on a cloud, grinning like Garfield, the cat.

Then Nick's tone changed as his voice turned serious. "I'm concerned how this will affect the close relationship I've built with my kids." He took a deep breath, almost a sigh. "I don't want my relationship with them to change."

"Nick, come on. When a new person enters the picture, it alters the family dynamic. But, that's not necessarily negative. It could turn out quite positive."

"True," he replied, but with hesitation. "My girlfriend's closer in age to my kids than to myself, by a few years."

His children and my own were in the difficult teens. I ran my hand through my hair, pausing to do the math, and concluded a difference of fifteen to twenty years between he and his girlfriend. In my children's life their father had also married a woman closer in age to our children than to their father. I chose understatement. "The relationship may call for some adjustment."

His voice took on a different tone, as though his heart and soul spoke. "Oh, dear God, I don't want to lose them."

"You're a caring father. Why would you think you would?" When he answered, his voice was barely audible. I strained to hear him.

"I just fear my 'me first' propensity. Every time in life I selfishly, impulsively put my wants ahead of those I love, life always teaches me a harsh lesson."

I speculated, "Maybe this is a test to see what you've learned?" Then I added lightly, "So you can graduate."

He chuckled, "Perhaps. You think that's it?"

"I don't know, Nick, but, as far as the age difference, I don't think love knows about age, or race, or religion. Love just is."

"Good point," he said in his marvelously expressive voice.

"If you two really love each other, your kids will sense it, be drawn to it, and to her."

"Ah-ha. That's true." His tone brightened considerably. "Happiness is contagious, isn't it?" He again sounded optimistic. "You know, I firmly believe God intended humans to be happy, including myself." Nick cleared his throat as if pushing away his reservations. "Which is why it's likely I have found my soul mate."

I liked his confidence. I thought his new amour represented dreams we all dream. I wondered about the mystery woman. With his attraction to tall, beautiful blondes, I imagined she possessed the sexy allure of Marilyn Monroe or the radiance of Julia Roberts. He's a terrific guy; I assumed she was likewise. Then he interrupted my thoughts.

"What do you think," he asked, "you're wise."

"Me?" I was surprised he thought so. In fact, for years I'd been teased that I couldn't think, but if I had a brain then I could.

"I'm flattered you think so," I replied. "I don't know quite what to say." I paused, my eyes searching the ceiling as if I'd find the answers I sought. Finally, I said, "Nick, I can't comment about another woman, or someone I've never met. Besides," I laughed, "you're asking me serious questions before I've had my morning coffee."

"That's it?" he retorted jovially pushing me for any further impressions.

I knew from his tone he wasn't settling for "no comment." In the background I could hear his footsteps as he paced the tile floor of his home. "But," he persisted, "what do your instincts tell you?"

Feeling on the spot, I took a deep breath before replying. In the back of my mind something didn't fit. "Nick, I hear you say you're thrilled one moment then riddled with doubt the next. No harm in listening to that still quiet voice before jumping in too deep, too fast?"

He groaned. Nick was a man of action, always in a hurry, and a tad impatient. "You're no help," he teased.

"Sorry," then I added, "Nick, I'm really happy for you. Follow your heart."

In the next second Nick's voice rang with excitement. "There's the door bell. She's here!"

Chapter 2

DISCOVERING THE OAK TREE

After talking with Nick, I was too wide-awake to return to sleep, so I threw back the bedcovers and quickly got dressed. After a cup of hot coffee, I headed for a trail in a nearby canyon in the Santa Monica Mountains. In the area, hundreds of centuries-old oak trees grow. The oaks are fed by water that seeps down through the mountains that run along the coastline. To reach the trail, I decided to cut through the greenbelt that runs through a residential area.

As I walked along, I had a premonition Nick and I would converse little in the months ahead. I'd sorely miss our talks. He made me think and had helped me learn much about Native American philosophy. These thoughts were filling my mind when I turned a corner and became distracted by the loud clamor of a flock of crows. Their noise was deafening. So many crows filled the branches of a tree that at first the tree looked black. I'd never seen such a sight. I wondered why the crows were making such a commotion. Then my dog at the time, a zany Airedale named Ruggles, raced toward the tree. I ducked when hundreds of crows flew toward me in a dramatic, noisy flapping of wings.

Mediterranean cork oak, Westlake Village, CA

Out of curiosity, I approached the oak, wondering what drew the crows to the tree. One huge crow remained. He jumped down onto a branch so that we were less than two feet apart. Crow's black iridescent feathers shimmered in the sunlight revealing every imaginable color. Like other crows he had one crossed eye, the other straight. Crow's straight eye bore into mine. Unafraid, I stared back. Behind Crow's eyes I sensed a higher order and hundreds of secrets about the earth. I thought society ought to expand the phrase "wise old owl" to include crows. It was as if Crow saw the outside world with his straight eye and looked inward with his crossed eye.

I nodded slightly, yet respectfully, in greeting. After we stopped appraising each other, Crow turned. He ambled along the branch toward the tree's main trunk, while looking over his shoulder to see if my eyes followed him. When he got to the trunk, he contemplated the bark. I thought, "Okay, I'll play along." When I gazed at the bark, I realized I'd never seen a tree like it.

For starters, the bark was very textured and contained three distinct colors. The raised bark consisted of undulating ribbons of pale silvery gray beside rows of dark charcoal. In the bark's deepest furrows ran ribbons of rusty-red. I checked the leaf, and clearly it resembled an oak.

The oak grew close to a tall, white stucco wall that enclosed a home and garden in privacy. The cork oak's canopy formed a forty-foot wide umbrella. The tree's size led me to conclude that landscapers had planted the oak decades ago when the greenbelt was first laid out. However, the oak had a problem. It didn't stand up straight. In fact, the oak tilted backwards at a thirty-degree angle toward the wall as if imitating the Leaning Tower of Pisa.

I rested the full length of my body against the massive tilting trunk to look up through the branches at a bright blue, cloudless sky. While admiring the pattern the branches made, I noticed a tiny twig had started on a huge, bare limb. The twig appeared so small, so irrelevant in relation to the enormous limb. I contemplated the potential within the little twig, wondering how big it would eventually grow, but I knew that the tree would determine its own growth. It was as impossible to accurately predict the twig's future size as it was impossible to predict what course Nick's new relationship would take. However, I had a feeling that although it was still winter, spring had already arrived for Nick and his girlfriend.

Thinking of Nick's relationship with his children, I noticed some branches intersected at right angles, each branch going off in a different direction. If Nick's worst fears came true, if he and his children grew apart as they grew up and left home, the potential estrangement wasn't necessarily permanent. The tree's persistent growth demonstrated that living things continually change. A tree's roots fix a tree in place, but its top branches remain supple, adaptable to the winds of change, as does the human mind and spirit. He and his children could find a way to grow close again. A tree—like love itself—continually creates anew, and he and his children could, too.

As the months passed, when I visited the oak grove I often saw in a flash how much everything on earth is related and interconnected. Back home, I'd spend hours mulling over what I'd glimpsed so briefly in the oak grove. By honoring the new insights, my senses awakened along with increased intuitive awareness. I began to understand that a loving God filled nature with knowledge and lessons for life. For thousands of years, millions of people would not have access to books, but instead could look to nature for wisdom. So the information that would help all of us live fully, joyfully, and peacefully was placed in nature where we could find it, if we looked.

In the tranquility of the oak grove, I began to understand that simple truths are waiting in nature for anyone to discover. When we physically go out into a garden, park, or the wilds, we can reconnect with nature. After relaxing to quiet the mind, we can enter the silence where we find the answers to personal questions.

Chapter 3

THE STANDING ONES

When I returned home, I did some research and learned the tree where the crows congregated was a Mediterranean cork oak that grew within a grove of ancient and majestic coast live oaks. I had never noticed the cork oak until the crows practically covered its branches. Hundreds of times I'd passed that tree, but it was simply another tree, the way a stranger is another face in the crowd. My attention had always focused on the more impressive native California oaks that surrounded it. By delegating the cork oak to a minor role, I judged it inconsequential. Certainly not something I had time to examine, too much to do. I never dreamt of the lessons to be learned within the circle of its branches. Before long I sensed there was something I knew intimately about trees, although that something remained illusive until three incidents occurred.

While at home, sitting at my desk, I glanced at a small resin sculpture my son had given me. It was a sculpture of a woman's face carved into a tree branch. Her long flowing hair fell in gentle curves that followed the graining in the wood. Since the beginning of time artists have shown a propensity to carve human images out of and into wood. On an intuitive level, artists sensed a fundamental connection between people and trees.

Shortly thereafter, I looked up a word in the dictionary and the page opened to a tiny illustration. At first glance, I mistook the picture for a cross-section of a human leg bone, but under the drawing the words, "cross-section of a tree limb," caught my attention. I was startled. Then I recalled that the earth is built with only a few basic elements in innumerable combinations, much like ten numbers make up all telephone numbers. The basic elements of life are carbon, hydrogen, nitrogen, and oxygen. The four elements are the basis of air, oil, water, plants, stars and humankind. As different as the billions of us on earth are, we all basically have the same DNA. I became intrigued with the idea that a limited number of principles might also be the building blocks of creation.

I wasn't in the habit of thinking that trees and people had much in common. We were separate, no connection, or so we'd been taught. By defining the differences that separate us from nature and each other, we've created loneliness. We are never alone. Everything and everyone is connected in the web of life.

Then I remembered author Lynn V. Andrews, in one of her books, had referred to people and trees as the earth's only *Standing Ones*.

Fascinated by the idea, for the next few weeks, I thought about what we have in common with trees beyond our shared need for the oxygen trees release and their need for the carbon dioxide we exhale. Eventually, I made a list of the qualities we share and here it is:

1. Each "seed" holds a powerhouse of potential and promise. The "seed" gets planted by impregnating the place where its life begins, be it mother earth or a mother's womb.
2. Tender young saplings and kids grow up fast then growth starts to slow with age.
3. In general, the younger the tree, or person, the greater the flexibility.

4. People and trees desire to grow straight rather than crooked. If either grows aslant, there's usually an attempt to straighten out, and most do so with time.
5. Both trees and people are highly influenced by the conditions or environment in which they're raised.
6. Although trees have much less control over their lives than we do, they share with us the great gift of determining when and how to "branch out" into the world.
7. In order to remain healthy, both of us need plenty of sunshine, water and nutrients.
8. Both people and trees make their own food.
9. We both have a built-in will to stay alive, to keep growing, and not give up.
10. All through life, mighty winds—both gentle and stormy—shape and give us character.
11. The older the bark, like the skin on the body, the deeper the furrows.
12. As trees and people age, there's a tendency to spread out, gaining girth. Some become more rigid, while others grow stately with age.
13. Trees are like the opposite sex; life becomes a desert without them.
14. A tree is like one's childhood home since it can offer shelter from the rain and sun.
15. In both the heart of a tree where the graining begins, and the heart of a person, there is enduring beauty.

CHAPTER 4

WILD THINGS

In general, people and trees begin life either planned or unplanned. In the planned category there are Christmas tree farms, orchards and formal gardens that are planned as purposely as a couple that conceives through fertility drugs or in vetro fertilization. Yet, for all the planning humans do, most life starts as a wild thing.

At night, in bed, a couple snuggles under the covers. They become inspired. Things heat up. Of the millions of sperm released, only a miniscule fraction will make it up the fallopian tube, but only one sperm needs to join a woman's egg and fertilize it for a new life to begin.

Outdoors, under the stars, a similar happening occurs as in the bedroom, but not nearly as fast. The male-like tree, erect, hard, and strong also releases thousands of seeds. Each tree has its own DNA as does each man. The maple tree has winged seeds borne in pairs. As the winged seeds leave the tree, they float down to the ground in a primal dance of life. The winged maple seeds turn and twirl as if winding down an invisible spiral staircase that imitates our own DNA's double helix. Although a multitude of maple seeds often cover the ground, so few seeds will sprout that it gives new meaning to the Biblical saying, "Many are called, but few are chosen."

After the seeds are on the ground, the maple tree sheds

its leaves as quickly as the couple removed their garments in the bedroom. Then the rains come. As in foreplay, the rain wets the brittle leaves, moistening, softening them until the leaves become like a down blanket that covers and warms the seeds. Under the composting leaves, heat is generated, encouraging the seed's germination.

Inside and outside the house, when life starts both beginnings are humble, silent. Although no trumpets blare, that spark of the Divine in every seed and human being is present and awesome.

Enormous changes begin to take place. Over the winter the leaves that were composting have turned into a layer of fresh soil. The new soil envelops the seed, protecting it in much the same way a woman's muscles surround and support the fetus in her uterus.

In the garden, the maple seed has developed a root hair that connects it to its mother, the earth, in the same way a placenta connects the fetus to its mother.

Inside the uterus and inside the seed casing the new life doubles and redoubles at an amazing rate. The remarkable activity that's taking place remains invisible for weeks, until the life inside is bulging outwardly, growing too large for containment.

When its time for birth an opening at one end of the seed casing begins to dilate, extending wider day by day, making it easier for the tip of a delicate leaf to peek out. At birth, an infant also emerges through a dilated canal. In order to pass through the opening the tender stem of a seedling is as soft and pliable as the bones of a newborn baby. Within the first year, the trunk of the tiny seedling will harden to support the sapling, in the same way an infant's bones calcify after birth to support the body.

As the months turn into years, the maple will grow tall and strong. The maple's taproot will reach deep into the earth as the taproot seeks underground streams. In a similar way as we grow into adulthood, the dreams that come at night tap deep into the subconscious where wisdom flows from a sacred well.

Coast live oaks on hillside, Malibu Canyon, CA

When we stop to admire a tree, wild or planted, on some level we are connecting, not only to our own humble beginnings, but also to the splendor within ourselves.

Chapter 5

LIFE IN A NUTSHELL

A blue and gray scrub jay carefully selected acorns from under a nearby oak. With the cock of his head the jay weighed each acorn in his beak, determining which acorn had been eaten by worms, which to bury, which to eat, and which to leave behind. Those acorns he saved for himself, he hid under the ground cover. Every fourth acorn, he systematically planted on the hillside. Using his beak as a hammer, he pushed the acorn down until soil covered it. In the process, his beak drilled a tiny hole in the top of the acorn, a birth canal so to speak. In time the hole would make it easier for a fragile leaf to emerge should the acorn take root.

Everyday, the jay worked to give back to the earth so that oaks would continue. What a miracle of planning and cooperation his efforts demonstrated, and a tribute to the ability for life to renew itself.

One of nature's building blocks is for living things to go through a dark narrow canal and move toward the light that offers a brighter future with more freedom. For an acorn to become an oak, a tightly furled leaf must get through the pin size opening the jay helped make, just as an infant must struggle to get through the birth canal. Birth is never easy, nor does it come with guarantees. When we were babies in the womb, the contractions must have made it seem like the walls of the uterus were crashing down upon us. Staying in the womb was no longer an option. Whatever waited outside, leaving the womb was worth the struggle.

We left the cramped, dark interior of the womb to labor—yes, as infants we labored too—sometimes for hours because the urge to live was so great. We painfully pushed, then pushed some more, just to get to the light at the end of the birth canal. After we came out, we opened our eyes briefly, just a slit, then shut our eyes from the glare of bright lights. We became aware of the silence when the reassuring beat of our mother's heart ceased along with the comforting sounds of her body's gurgling liquids that sounded like a mountain stream tumbling over rocks. Hospital sounds replaced the familiar, naturally soothing noises. With so much to take in, we held our breath before we took the first breath alone, independent of our mother.

For the next several months we were cuddled, diapered, fed and burped. Then we longed to leave the comfort of our mother's arms, and began to crawl. Desiring still more freedom, we sought greater mobility. We took that first step, fell several times, and tried again and again, until we walked. Finally, the day came when we ran across the grass, arms outstretched, smiling, feeling the free wind on our face.

The drive to become transformed by moving from confinement to a freer state is shared by all that lives. It happens over and over throughout life. We are not alone in this pattern.

A baby bird pecks at its shell. His beak becomes calcified, hard enough to poke a tiny hole. More pecking occurs. One day the nearly exhausted little bird writhes and wiggles until its head pokes through the larger opening it has finally made. Pushing upward, its shoulders crack the shell further. Then, like us as infants, the bird steps into a world without guarantees. The day will come when the bird flaps its wings, leaves the nest, and hopes to land on the nearest branch.

A caterpillar repeats the pattern when it spins a dark chrysalis that holds the promise that one day a butterfly will work its way out and fly away in hopeful search of the sweetest flowers.

On a plum tree, the buds wait for spring when the green sheathing dilates allowing the tips of the pink buds to poke into the sunshine. The blossoms open gloriously, but without guarantees that bees will come to them.

The passage from darkness to light occurs daily. We see darkness descend on us every evening knowing that the night gives birth to the dawn of a new day.

In sea floors across the globe, molten lava waits below the surface. The hot lava bubbles and pushes to escape from the dark center of the earth. Up through the mountain the lava travels then erupts into the air and falls, flowing freely across the land.

In Yellowstone National Park, Old Faithful sends its geyser of steam out from the earth high into the air.

Even for a movie to be made, the film must leave the dark canister, go into a camera, and become exposed to light.

The urge to escape confinement doesn't end at birth, but continues every day of every year. It may be as simple as getting done with a time-consuming project, studying for finals, or cleaning one's room before going out to play.

Then there are other more serious life challenges wherein we're unable to move forward. We feel stuck—but life like birth—requires that we work our way through the situation, if we want to be freer and able to continue to grow as a person. Just as breech babies seem reluctant to leave the womb, some of us aren't ready or feel unable to move forward. But, like a breech baby, at some time the infant has to turn itself around, or have someone help turn the baby around before it can travel through the birth canal. Like leaving the womb, living and loving are journeys into the unknown where there are no guarantees, but a passage worth taking.

After we take action to make a dream a reality, we are rewarded with enormous growth. Even if our first attempt does not work, on earth failure is usually not final. All the essentials we need to live are already inside of us, just as everything needed to become an oak is in an acorn. Although life has risks, earth is a nurturing place that was lovingly designed to provide for all. "And God saw all that he had made, and it was good."

The next day, the scrub jay was back at work living and planning for the future. Then he flew away. As I watched him fly into the sky, a strange thought came to me. Since it's a universal pattern that "birth" requires time and effort to go through a dark tunnel to move into the light, could black holes be passageways into other worlds?

Scientists will answer that question. But for now, holding an acorn in the palm of my hand means seeing the story of life's labors in a nutshell.

Chapter 6

SOUL MATES

Within a few months, Nick, the hunk in the first chapter, and his girlfriend became engaged. Shortly afterward, I was talking with Jennifer, my close friend and fellow writer, about the whole idea of soul mates. Somewhere in the conversation, I remarked that soul mates counterpoint in nature would relate to two trees that grow together forming one tree. Jennifer's eyes lit up. She related the love story of Baucis and Philemon, an elderly married couple, who turned into conjoined trees upon their death. Ovid, a first century Roman poet, immortalized the mythic tale that has lasted over two thousand years.

In the story, vacationing gods Zeus and Hermes traveled the country disguised in plain clothing. As evening approached the gods needed food and a place to sleep. Everyone they asked, even the wealthy households, turned them away. Tired and hungry, they approached Baucis and Philemon's humble cottage. Although they didn't recognize the gods, it didn't matter. They welcomed Zeus and Hermes graciously, fed them what little food they had, and gave them their own bed for the night. Baucis and Philemon's openhearted generosity impressed the gods.

Since Zeus, the greatest god on Olympus, and Hermes were turned away by wealthier households that had known them, they became vexed by the others' disrespect, and being vengeful, they decided to punish the others. The gods flooded the whole region, transforming it into a boggy swamp, except for a hilltop near the couple's small cottage, on which they built a great marble temple with beautiful gardens.

When their wrath was spent, the gods asked the couple if they had any special wishes.

Baucis and Philemon chose practical considerations namely, a place to live since their cottage was a bit damp. The couple asked if they could serve the gods as temple custodians. They also wished to die at the same time for it would prove too painful to live without each other.

The gods agreed.

When the couple grew very old, they died peacefully with their arms entwined. Where they were buried two trees sprung from the earth, an oak and a Linden (lime) tree. The oak and Linden grew together as entwined as Baucis and Philemon were in life and in death. Whenever people visited the temple, they marveled at the sight of two totally different trees growing as one.

In nature, two trees of the same type can appear as one tree, especially if the bark encloses both trunks. If that tree were cut down, the stump would reveal two separate "hearts." Surrounding the "hearts" (which are the original seeds) there are two separate and distinct sets of growth rings, evidence that even though the two trees grew as one tree, each tree retained its own identity. That's nature's wisdom reminding us not to lose our individuality in a relationship. Nature emphasizes the importance of valuing and retaining the unique, yet separate, gifts each person brings to a relationship.

The fact Baucis became an oak and Philemon a Linden tree holds more symbolism. Oak trees, which normally live for hundreds of years, represent slow growth that produces a dense hardwood that's strong and durable. In Europe many centuries-old stone manors and pubs are still supported by their original oak beams. Thus, oaks denote permanence. In Roman mythology, the Linden tree symbolized conjugal love. It has heart-shaped leaves and clusters of fragrant spring flowers that draw thousands of bees. The honey the bees make from the Linden tree is white in color and high in quality.

The security of a dependable, permanent love speaks to our heart's longing to find a love that refuses to die. Perhaps the longing for a forever love represents something deeper and more elemental: the soul's desire to return to the source of unconditional love. And, so our human desire for genuine love conjures images of a blissful, heavenly state. Yet, dreams of endless sunny days and nights filled with fireworks that light the bedroom seem unlikely when we look at nature.

There's nothing in nature that suggests unending earthly bliss. Nature remains in a state of flux as it adapts to constant change and seeks balance by the juggling of opposites: sweltering summers and freezing winters, calm and gale force winds, plus periods of flooding followed by drought. In nature and in our lives we have the same task, to create balance and order out of constant imbalance and disorder. During the 1995 Northridge, California earthquake, nature didn't discriminate in the damage created. The ground shook throughout the region. It collapsed sprawling hilltop homes of the rich and famous and low income housing with equal forcefulness. Nature has an essential equality to it. Therefore, it seems unlikely that soul mates receive special treatment that shelters them from life's struggles.

Soul mates are still human beings, and as such they cope with mortgages, stresses at work, illnesses and raising children the same as their neighbors. Every year the experiences of earth—the laughter and tears, love and indifference, sickness and health, along with triumphs and tragedies—are common to every one of us.

However, soul mates do have two special qualities found in Ovid's mythic tale. Philemon, as a sturdy oak tree, provided a reminder that despite the hard times their poverty gave them, their love weathered life's storms. In nature, a severe storm stimulates trees to dig deeper, and grow stronger roots. When earthquakes shake tress to their very roots—as happened many times to so many ancient oak trees that dominate the California landscape—the quakes support Ovid's choice of the oak as a symbol of lasting love.

Ovid's selection of a Linden tree to represent Baucis is equally significant. The hunger soul mates satisfy is the nourishment our belief in love, the most beautiful, powerful force in the universe. The delicious honey that comes from the Linden tree or the sweet fruit from fruit trees provide food for the body just as love provides food for the heart and spirit.

Finding that certain someone, who aids in our own evolvement and we in his or hers, is a most precious gift. It is a reward from the universe. If it takes a long time to find real love, take heart because in nature, it also takes several years for a fruit tree to produce a significant harvest. The tree first has to grow strong and mature. Even after a fruit tree reaches three to five years old, there's no guarantee it will yield a noticeable harvest, anymore than one's true love will arrive when we wish. Conditions must be right. A cold snap can limit the blossoms on a tree. A hale storm can strip the blossoms off the branches. After a tree blossoms, it needs the birds and bees to fertilize the blossoms before fruit can form. In a similar way, attracting that special person is often a reward that follows years of healthy positive growing conditions.

When we find true love, it's possible that more is expected from such a love than from other couples. In nature, fruit trees must do more than other types of trees. In addition to holding the soil in place, providing oxygen for the atmosphere, and shade to cool the earth in summer, fruit trees added function is to produce fresh fruit that benefits other creatures. Soul mates probably have a purpose beyond caring deeply, passionately for each other. Nature shows us that just as a fruit tree shares its fruit freely, soul mates must share their love with others, their family, their children, who in turn pass the capacity to love onto their children and grandchildren then into the community.

Failure to use the gift of love may result in serious omissions and deep regrets for neglected opportunities. If a couple hoards their "harvest," by refusing to share the "fruits" of their love, in other words, by not expanding their love outward beyond themselves, the fruit of their loving might fall to the ground to rot and decay. Nature shows us what happens in an orchard when fruit goes to waste. Flies swarm to the decaying fruit, and eventually the air near the fruit tree smells like vinegar.

Baucis and Philemon, as an oak and Linden tree, together symbolized that the great love often associated with soul mates, not only endures, but the couple also shares their love. Real love is never miserly or limiting, nor is it controlling, demanding, confining, jealous or possessive. Love is kind, expansive, constant and generous; it keeps growing, becoming stronger and more bountiful year after year.

Chapter 7

THE BONSAI

As I stood on the patio, my line of sight encompassed two trees—a young wild oak that grew on a hillside beyond the wrought iron fence, and a bonsai my friend and I purchased. We intended to look after the bonsai as we looked after each other. We did for awhile. After a couple months, he thought the bonsai needed more sunlight. So just before he left, he put it in the garden without mentioning he'd done so. He was quite fond of the bonsai, so when I noticed it was gone, I thought possibly he'd taken it home with him. I made a mental note to ask.

Several days later, when we spoke on the phone, he said the bonsai was in the backyard. When I found it, the bonsai desperately needed water. It's once shiny green needles had yellowed from our take-it-for-granted attitude when each assumed the other was looking after the bonsai. To restore the bonsai to a semblance of its former glory took hours of using tweezers to patiently remove the dead needles. Afterward, the little pine appeared rather sparse and sad.

When he came to the house the next day, he took the bonsai from the filtered sun on the patio to deep shade inside the house in compensation for the days when it had had too much sun.

The bonsai got confused.

I think what happened to the bonsai isn't much different from relationships since the latter also require tender loving care to revive them when they begin to wither. Perhaps this is because relationships embody similar qualities to potted plants. For instance, a close relationship may carry an implied agreement to live within certain boundaries, the way a plant lives within the vessel in which it's contained. There are limits to the dimensions of the planter, just as there are limits as to what is acceptable to both people. The bonsai isn't expected to spread its roots in a way that causes the pot to crack, or to stick its roots in another pot.

Single persons are similar to the oak beyond the fence since both are responsible for meeting their own needs. On the other hand, the bonsai lost much of its independence by being contained in a flowerpot. Implied within the connection between gardener and bonsai, there are—as in other close relationships—certain expectations. It might prove prudent to enter relationships with "how to care for" instructions attached to our hearts, since hearts break the same as clay pots.

The care instructions might go like this:

You are entrusted with the bonsai as a symbol of how much we care for each other. If you wish it to grow fuller, and be all you envision it can become, treat it, as it needs to be treated. In order for it to flourish in its container, keep it moist, remembering that not enough nurturing water causes it to shrivel and dry up, whereas too much water smothers it.

The bonsai needs the reflected light of the sun as people need to bask in the reflected twinkle in the eyes of the people they love. If expressions of appreciation are closeted, darkness will surround the tree then disease will follow. Provide the balanced light the bonsai needs to thrive. Let the bonsai do the same for you. Don't over-fertilize it, forcing it to grow, getting all swelled up until it overshadows your light. It doesn't want to become greater than or less than you. It desires to give mutual respect to creatures great and small to the best it is able at each moment in time.

Talk to the tree and tell it what's on your mind. This is a listening tree, not a judging tree. Reward it with affirmations when it makes you happy. It is not a guessing tree. A smile or hug does more to reinforce learning your desires than silence.

You may feel a need to complain about it occasionally by telling it what you don't like or wish it would do. But leave the final choice for change up to the plant. It alone must do its own growing. Do not cheat it of its Divine purpose and potential by wanting it to be less than true to itself. It's unfair to resent its inability to grow to your specifications.

The bonsai will honor the freedom you give it. In time its inherent beauty will reveal itself. While freedom may stir insecurity, remember that the tree has roots. It doesn't fly away.

People and plants need a balanced environment to thrive and give and receive pleasure. Even though you call it your tree, it is only yours as long as it can be with you.

Each person and plant is different, but each is a reflection of the Divine. May we always treasure the differences for they need not divide us as much as they can complete us. All living things learn from each other, and need each other.

The bonsai is much more than wood and greenery, just as you and I are more than flesh and bones.

CHAPTER 8

FOUR DIRECTIONS

In a former home, outside the kitchen window, there was a relatively steep slope covered in ferns. Atop the grade, on the property line, stood a wrought iron fence, and slightly below the fence grew a small acacia tree. In spring the acacia brimmed with yellow blossoms. The following year new neighbors planted a little plum tree on their side of the fence. As the years past the plum and acacia trees grew large. Each vied for space to spread their branches. The plum tree, with the best plums I've ever tasted, won. The acacia grew lopsided, unbalanced, its growth lush on the downhill side, yet stunted on the opposite side that faced our neighbor's house.

One October morning, mighty Santa Ana winds arrived in the Los Angeles basin. When I came downstairs to make coffee, I found the twenty-foot acacia tree in the kitchen. Fortunately, the sliding glass door was not latched and the tree merely pushed the door open rather than crashing through it. Across the kitchen table lay the acacia's upper branches, looking like an enormous, out of scale centerpiece.

The acacia, if its growth were balanced, wouldn't have ended up in our kitchen. Trees need the steadiness that comes from their branches growing in four directions so they can stand up to mighty winds. Like you and I, trees are living entities and they acquire strength and stability when their growth makes them well-rounded. We also have "four directions" or sides to our humanness: emotional, intellectual, physical, and spiritual.

Throughout our lives we're subject to the winds of change that bring storms into our lives. Storms can topple even the highest, mightiest among us. If one is to stand tall and strong, fortified for what the future brings, then the lesson of the acacia is recognizing that one's emotional, intellectual, physical, and spiritual sides need to be honored so that one can stay sufficiently centered when times are tough.

There is no co-dependency among trees. The acacia could not expect the plumb tree to grow its missing branches so the acacia would become "whole." The acacia became weak when it stopped growing in a balanced manner.

When I thought of the connection between the acacia and plum tree, I thought the importance of the "four directions" also applies to interpersonal relationships. If one of the directions is missing, issues often arise. Suppose a person is attracted to someone, who fulfills his or her sexual needs, but doesn't meet his or her emotional needs. After awhile the best sex in the world isn't enough. The lover that met the sexual needs gets confused, and says. "Hey, this isn't fair, you said I was incredible. Why are you pulling away? What's wrong?" One of them answers, "I need more."

Suppose a couple has great sex, but intellectually they are so far apart that through the years loneliness sets in, then a hunger develops to alleviate mental boredom. Perhaps career building consumes one partner while the other one spends the days learning about emotions by raising their children. The world each person visits on a daily basis has slight reality in the other person's perspective. The couple's growth is weighted in different directions, each needing to nurture the area that is stunted.

This direction idea works in other ways. Say a married couple shares intellectual interests, great friends and stimulating conversations, but they're not lovers. After a while, a physical hunger sets in. Discussing all ten books on the best-seller list, or other areas of interest, doesn't cut it. In time, physical needs, the joy of touching and being touched, rise to the surface and an undeniable physical yearning occurs.

Perhaps a partner who hasn't said a prayer in years starts to yearn to grow spiritually, much the same as a tree must reach for the light. But, if the other partner can't relate, a sense of isolation grows in the heart of the person seeking to evolve spiritually.

Many of us meet someone, fall in love, and life takes on a rosy glow. It's easy to ignore the signs, and think that we can live without one of the "needs" being acknowledged or shared. We can, for many years even, but then, like the acacia that didn't grow fully, a wind stirs the neglected need in ways that can no longer be denied. Taking action to fulfill an unfulfilled need is often perceived as a threat to the relationship. In reality, ignoring the need is ultimately just as great a threat.

When we pay no heed to nature's wisdom and ignore our inner voice, discontent eventually results, and that lack of ease can lead to dis-ease. When we listen to nature's guidance to become strong by growing in four directions, we are rewarded with a sense of wellbeing. We become as well-rounded as a sturdy tree whose branches spread out in a circle. It is then that we attract another person who is as equally well-rounded. Then we are more likely to grow side by side through the years. Such couples are unlike the plum and acacia trees whose time together was short-lived.

Chapter 9

WIND STORM

We tend to think of nature as separate from ourselves, but we share more than is visible on the surface. This is the story of Brad, the father of two young children. He and his wife were legally separated when he met Lisa. It was a wild and passionate affair that went on for five years. During that time, he introduced her to his world. They traveled to Paris where they stayed in his apartment with its view of the Eiffel Tower. In the South Pacific, they watched stunning sunsets while sipping tall tropical drinks. On a trip to New York, they spent a week in Manhattan seeing Broadway shows and dining at the finest restaurants. Oftentimes, the happiness they found together caused total strangers to comment on the joy they radiated.

Brad became quite satisfied with the status quo of the relationship. His finances remained intact, as did his friendly relationship with his wife and their children.

Lisa suspected her relationship with Brad would not progress further because he made no move to get a divorce. As her biological clock kept ticking, she grew increasingly frustrated and annoyed with herself for staying with him. When a new guy caught her eye, she gained the impetus for change, and she ended the relationship with Brad.

Not surprisingly, it devastated Brad. With a desire to prove his sincerity, he started divorce proceedings shortly after their breakup. Lisa's relationship with the new guy only lasted a short time. Brad seized the opportunity to win her back.

One might expect a happy ending, but that was not the case. The more he transformed himself to meet her desires, the more he came up empty-handed. The more he gave, the more she reminded him that he had hurt her. His not wanting to marry her sooner made her feel like a mistress. As time progressed and they talked infinitum, memories of her father increasingly crept into their conversations. When she was a child, her father had had a mistress, and he left the family. She urged Brad to go into therapy to find out why he didn't want to make a commitment earlier in their relationship. He did go to therapy, and learned much about himself.

When Brad wanted to leave the baggage of the past behind so he and Lisa could move forward together, he presented her with a diamond engagement ring at their favorite restaurant. She ran away, out the door, and took a cab home. During the next week, she didn't answer his calls, or her door. He sent her flowers.

"What do you make of it?" Brad asked me.

Outside the Santa Ana winds rocked the towering eucalyptus. Swaying branches bent low, scratching against the huge window overlooking the garden. Knowing the ups and downs of their relationship, my thoughts shifted from his point of view to hers. As I tried to understand, I watched a ten-foot limb snap off and sail through the air. It landed in a nearby flowerbed, sticking out of the ground like a spear.

His eyes followed mine to the scene outside.

"Look at how the trees move," I said. "Seventy-mile-an-hour winds are pushing the tops of the trees while the trunks don't move an inch." He nodded. "Reminds me of your relationship. You're the top. She's the trunk. You've changed as requested, but she has not budged." With hesitation, I asked, "Is she attached to discontent?"

His dark brown eyes narrowed. "Yes," he said, "but I believe she'll lose that quality with the life I'd give her."

"Your childhood was incredibly blessed."

He nodded. "Hers was difficult."

"The differences in your backgrounds are as firmly rooted as those trees." Another branch sailed through the air. "A tree grows and stays healthy by getting rid of dead wood."

Brad set down his coffee cup and leaned over the sofa. His eyes scanned the broken branches littering the lawn. "What are you saying? That we need to get rid of the sad memories like the tree gets rid of old wood?"

"Yes, it's not healthy to live in the past," I replied. "Nature is pruning the trees, encouraging new growth. The new growth represents a new tomorrow." We stopped talking and listened as a powerful gust of wind rocked the house. The wind gave me an idea. "Hey, you asked why she ran away. Is it possible that by marrying you, she'd have to give up the dead wood?"

He pulled on his chin mulling over the idea.

"Brad, you're so positive. To marry you she'd need to give up distrusting men that she absorbed when her father left in childhood. Your proposal probably created a windstorm inside her. Maybe she ran away to go where the air is still."

"The eye of the hurricane. Humph," he said, "that makes sense."

"Right now she's comfortable where the air is stationary. She's not answering her door or her phone."

He frowned, "That's for sure."

"Perhaps she's not ready to give up the dark shadow that men are suspect, which her mother drummed into her. She's not alone. There are some men just as suspicious of women. Oprah Winfrey calls them the shadow ideas we carry around with us."

"What do I do?" he asked.

I shrugged. "Give her time. Deadwood doesn't form overnight, nor does it blow away fast. It takes a storm. You've created one. Ride it out." I patted his shoulder. "She wouldn't overreact, if she didn't care."

Although his mouth formed a slight smile, his voice was sad. "If she loved me, she'd want us to be together."

"Brad, it's probably not about you, but the childhood issues you stirred in her. Out of loyalty to her mother, she's learned to discount men and her stepmother, and now you're asking her to be a stepmother."

"Oh, God." He ran his hands down his face.

"Or, she may be afraid of commitment. Who knows? You haven't talked with her since she ran out of the restaurant."

"She may not know why she ran away." He leaned his head against the back of the sofa, closed his eyes, deep in thought.

"What are you thinking?"

He shook his head. "I don't know if I can live with always having a problem. You're right. It's deadwood. I want us to be happy. There's so much to enjoy."

During the next year, they saw each other, but he kept coming up empty-handed in terms of marriage.

In the process of getting a divorce, instead of giving his wife sole custody, he sought and got joint custody. Then he devoted more time to his children, especially his little daughter. He wanted to minimize the effect the divorce had on his children because he didn't want them to carry a hurt that didn't heal. Most of all, when his daughter grew up he wanted her to embrace love when she found it.

Chapter 10

THE SKATER'S ATTITUDE

The opposite of deadwood noted in the previous chapter is new growth. A symbol of new growth appeared the day I met an Asian boy while walking along the greenbelt. He appeared about nine years old. The boy came unsteadily toward me on roller skates. On his body he carried much extra weight, which combined with his irrepressible grin, reminded me of a smiling Buddha. I noticed the grass stains on his white tee shirt and a few bloody scratches on his arm and leg. I smiled and stepped aside as he approached.

His eyes twinkled. "Hi, how are you?" inquired the jolly stranger as he tried to maintain balance to avoid a bump in the uneven cement walkway.

"I'm fine," I replied, thinking there's something magical about this kid. "Are those new skates?"

His head bobbed affirmatively. "My sister got hers first," he replied, wobbling as he hit the bump. With his arms waving he tried hard not to fall. I reached out my hand. He took it and managed to realign his balance.

"Hurry up, don't stop," called a confident tall wisp of delicate beauty that I took for his older sister.

Smiling up at me, he continued. "I told my Mom I wanted to learn. Just got them," he added as he lumbered away breathing heavily. He might have opted for watching TV since it was easier. But, with a determined glint in his eyes, he glanced over his shoulder to see how close his sister was. She was closing in fast. She gracefully glided past us. In that moment, in the boy's eyes I could see the non-skater faded and a skater was born.

I recalled my first time on roller skates. The earth seemed to turn to Jell-O. Forward movement became precarious, like a baby's first steps. Success in skating comes from making order out of infinite possibilities such as when to move your feet and when to coast. Another requirement is a positive attitude, and this boy had determination to succeed along with humility to get up when he fell. With a smile and a wave, he skated past me.

He was skating better when a boy on a ten speed possibly imagining he was Lance Armstrong in the Tour de France came barreling around the corner. The skater met the pavement.

The cyclist hit the brakes. "I'm sorry," he called, immediately turning his bicycle around. He came up and inquired, "You okay?"

Too stunned to say anything since the cyclist didn't come within three feet of him, the skater accepted the cyclist's out stretched hand. It's harder to get up from a sitting position when you are overweight and wearing roller skates. The cyclist dug his heels into the grass and slid the skater off the concrete and onto the lawn area. When he got up, he looked to see if his sister had seen him fall. She hadn't. The two boys talked briefly then soon both went on their way.

A week later, the indomitable spirit of the skater shone as he flew by me. This time he had no grass stains on a pale blue T-shirt and radiated a smile of pure joy. "Hi," he said quickly, "I met you before."

"Yes," I replied, "and now you're up to freeway speed."

"Thanks. See you." He giggled as he continued on his way.

The typical tree gets only a small percentage of its nutrition from the soil. A high proportion of a tree's nourishment comes from the atmosphere as air, rain, and sunshine. The skates didn't make the skater. The jolly skater got his skill from high up, too, when the atmosphere his thoughts created became reality. Thought matters.

Chapter 11

LEAVES OF RED AND GREEN

On an autumn day shortly after the leaves were ablaze with fall colors, I was reminded of the value of being true to oneself and that perfection doesn't promise happiness.

The awareness occurred after I collected a bunch of bright red maple leaves. It surprised me to see bright green on the underside of the leaves—Christmas colors, a foreshadowing of the holiday that was only weeks away.

I put the vibrant leaves in a basket on the kitchen table. A few days later, the colorful leaves had lost their red and green hue and turned an ugly brownish-gray. I thought about the drastic change. Red and green are opposites on the color wheel. When opposite paint colors are mixed together, they neutralize each other then form gray.

The withered leaves reminded me of someone I deeply loved who in many ways was opposite to me. Where I was pastoral green, hopeful, and easy-going, he was peppery, ardent, continually racing-to-a-fire red. He lit a fire under me, which was beneficial, and I soothed his fiery nature. Our chemistry was much like the autumn leaves wherein two opposites formed an exhilarating, distinctive union.

Opposites can bring excitement to a relationship when their differences complement and balance both personalities.

Opposites also hold the potential for a third dimension, healthy tension for growth and prevention of boredom.

I believed neither of us desired the other to lose his or her individuality. Then one day he said that if I loved him, I'd do certain household tasks in certain ways. His requests were neither immoral nor illegal. The appeals had to do with perfection of matters concerning the home, and I found the requests a bit excessive. But, if stacking the magazines a certain way made him happy, fine. Then there were more requests that eventually became mountains of exactness.

As a couple, I lost my quiet tranquil green hue when I submitted to meeting requests for perfection, when I was clearly neither perfect, nor a perfectionist. Blinded by love and willing to keep the peace, I didn't realize I was often on edge, worrying that something would make him irritated. At the time, I didn't have the good sense to know that if he cherished me he would not try to tame or change me, but he'd treasure all parts of me, including the imperfect parts. When people—like the maple leaves in the basket—lose the intensity that defines their exquisite individuality, they minimize celebrating who they are. When people loose their vibrancy, they fade like the dying leaves in the basket. I eventually became the same blah color as the lifeless leaves. When I'd submerged the parts of me that made me unique, I was no longer interesting even to myself.

It is presently several years later, and I'm no longer with him. Recently, when autumn's leaves turned bright and colorful, memories of those days returned. I started thinking of the pressure and futility of trying to achieve perfection which is not about perfection at all, but control issues.

My thoughts were interrupted by the TV newscaster, reporting the start of a search in Oregon for the perfect Christmas tree to be used in 2002 in the White House. I shook my head at the thought that among the millions of

Douglas Fir trees in the state of Oregon foresters needed *two years* to find the "perfect" fir tree. Deep inside I know there are no perfect trees. There is always some flaw with a crooked branch or a hole in a tree's silhouette that disturbs the symmetry. We are much the same for it is our imperfection that connects us to nature and gives us our individuality.

Then I recalled the beginning and ending lines of Joyce Kilmer's poem, *Trees:*

>I think that I shall never see
>A poem as lovely as a tree.
>
>Poems are made by fools like me
>But only God can make a tree.

I smiled knowing that God, by imbuing trees and humans with imperfections, didn't expect either to be perfect as a condition for being loveable and wondrous.

Douglas fir tree, Lake Oswego, OR

Chapter 12

KARMIC RELATIONSHIPS

People find love in one form or another. Or, more correctly, what seems like love based on what one thinks love is.

In the first chapter, Nick's voice rang with awe and wonder when he said, "The relationship is so karmic." He sensed that fate definitely had stepped in to bring him and his new love together. Mathematically, his exuberance was understandable. Considering that there are over six billion different people scattered throughout five continents, it is reasonable to conclude there is someone for each of us. And, Nick believed he had found the proverbial needle in a haystack.

Some people look for a lover that is as graceful as an elm, such as an athlete or dancer. Others know themselves well enough to know that, like the cottonwoods, they need to live near water and they shy away from those who must live like the Joshua Tree does in the Mojave Desert. Some people seek a lover as sweet as the sugar maples, while others want a love that's hot and spicy like the fruit of the pepper tree. Sometimes size matters since a person may want someone short like a dwarf tree or tall like the Redwoods. With the population increasing as people live much longer, more folks resemble the bristlecone pines in longevity. Many seniors find love again.

We all want to love and be loved in return. Yet, despite the earth's great population, the chances of finding that special person are reduced, considering many of us have neighbors on the next block that we've never met. The chance of two compatible people meeting when they are both available, living near each other, let alone falling in love, narrow the possibilities; it gives credence to the saying that there are no coincidences.

When we meet that someone we're unexplainably drawn to, we get excited at the possibility a meaningful relationship might develop. While I agree with Nick that karmic relationships are a special gift from the universe, I have a slightly different take on them. I regard karmic relationships as heaven sent learning experiences whose outcome can go one of two ways. Some relationships work, some don't. Those that sour seem stamped like dairy products with an expiration date that was overlooked.

Such relationships remind me of a tree I planted with great expectations, yet after a few years the tree withered. Couples know when their love is withering, although they never want the loving feelings to end. While their love fades, they keep hoping their love will sprout new growth to signal it still has life in it.

When a couple separates, each person goes through stages and pining may fill one's heart. In the next stage, we often get angry and feel stupid that we didn't see certain traits before we "planted" a love that wouldn't keep growing. We're amazed that someone who seemed so right could turn out so wrong. Nature encourages us to take heart and not beat ourselves up for not knowing the outcome in advance.

How are we to astutely discern the complexity of another human being when, in a seemingly simple act of identifying a leaf from a tree, we are easily misled? Shiny, dark green apricot leaves are readily mistaken for those of the poplar. A leaf from a cherry tree is dark green, oval with tiny teeth along the edge similar to an elm leaf. If you can't see the Sycamore's "camouflage pattern" on its trunk, but only see the leaf, a person can mistake the Sycamore leaf for that of the maple.

One day my friend, Arlene, who is part Cherokee, and I walked in the woods. She pointed out different edible plants of which I was unfamiliar. When I misread a hazelnut leaf for an alder, she said, "Turn the leaf over, if you're not sure. The underside, where the veins are, tells how the sap flows."

How many of us stop to analyze a relationship that seems a destined match? Few of us take a wait-and-see attitude until both have viewed each other's underside. A go-slow attitude proves difficult when we believe true love is knocking. Instead, we eagerly fling open the door to our hearts and invite love in. And, nine times out of ten, we are lured to that person by an incredibly strong physical attraction. If it wasn't for that strong physical attraction, we might have quickly turned away from the one person who will teach us an important life lesson.

Later when one or both decide it's best to move on, there's often an inevitable feeling that the relationship may have been some sort of Cosmic Joke. It wasn't.

Every relationship is sacred, but karmic learning relationships provide experiences that further the soul's growth, and, as such, they are a precious gift. Although such unions may bring emotional aches and financial loss, those relationships "light a fire under us," that cause us to change fast and often thoroughly. Perhaps their purpose is to make us more astute, more deeply aware of what we do and do not want in a partner, and the type of person who is better for us. Such self-knowledge is a reward because often it is by what we reject that we are most clearly defined. We emerge wiser, a more complex and interesting person. If the relationship ends, bless him or her for being your teacher and the hurt goes away faster.

Earth is a learning place where acquiring knowledge is facilitated by a body that lets us learn quickly because we can feel; the heart expands for joy, contracts with sadness.

In the same way that teachers give tests in school, the universe also tests us to see if we will make choices that are beneficial and loving to our soul. If we're slow learners, we'll get another chance in the next relationship to understand what we didn't fully realize the first time.

When going through a rough patch, nature provides two valuable lessons. The first is that, following a storm, plants dig their roots deeper into the earth and wrap them around rocks. In this way they gain stability to minimize their vulnerability against future storms. In a similar way, we reach deep inside ourselves to find out what it is about us that drew a person that is not apt to bring us happiness. We seek the stability of "rocks," such as astute friends and family.

The second example nature provides to get over a rough patch relates to the seasons. Summer represents the rosy, warm glow of the relationship. Fall relates to when the romance begins to change color and wither. In late autumn deciduous trees set their fading leaves free to fly away with the wind. By recycling dead leaves, nature shows us what to do. It takes every single leaf, like every moment spent together, and through the winter it forms compost that is as rich and useful to the future as wisdom. The tree's bare exposed branches resemble the way we feel after a break-up. This is normal. Winter can be cold and lonely. But, a tree without last year's leaves is lighter like we are without the stress of a difficult partnership. We don't have to carry around what's no longer needed.

By spring the hours of sunshine, like joy, gradually increase. In the trees—the other Standing Ones—nature shows us not to remain in the winter state too long. It takes about a month for every year a couple was together to get over the other person. No matter how bitter the winter, or painful the relationship's ending, every spring trees bloom again. Spring is a promise that's been kept since time memorial. The same thing happens to us because love, like a tree in spring, will bloom again, if we are willing.

Chapter 13

CHOICES

Along the greenbelt, there is a coastal live oak of such great height and width that the slow-growing species is easily five hundred years old. Its trunk stands so straight and sleek that it doesn't begin branching until it clears the chimney of a nearby single story house. I call the tree Big Billy.

The tree's enormous height contrasts with its miniature leaves. The leaves on coastal live oaks are typically small, one—to two-inches long, as dark as holly leaves, shiny, waxy, and cupped or spoon-shaped to save moisture by exposing less leaf surface to the drying effects of the sun.

On the towering trunk, there's a low spot where I've often seen a little girl raise her arm and give the tree a gentle pat on her way to school. In February, which is early spring in Los Angeles, on the spot where the girl's warm hand touched the bark, two little leaves emerged. Then a twig formed to support the leaves. By summer, the twig had forked and its off shoots added many more leaves. It progressed into a three-foot-long branch that stuck out on the side of the lofty trunk. The branch represented a change, a shift, and a new direction in the life of the ancient oak that started with the warmth of a little girl's affectionate touch. Only time would tell if the mini-branch would grow into a significant limb.

Valley oak, Westlake Village, CA

One day when the blue sky peaked through thousands of leaves on the oak's huge limbs high overhead, I became aware that massive limbs all begin the same way as the mini-branch started, a leaf at a time. In my mind's eye thousands of leaves on the mighty oak came to symbolize the countless choices that we make throughout life. Some options, like a leaf that never grows beyond a small twig, resemble momentary choices the results of which have a minor effect on who or what we become. Such selections as whether or not to watch television, rent a movie, or what to eat for dinner on a summer's eve are generally inconsequential over a lifetime. Whatever one ate on a Wednesday twelve months ago often remains forever beyond recollection. That's why life's minor decisions are similar to deciduous leaves or annuals in a garden, they fade by autumn. Winter rain and snow will turn once colorful leaves to black loamy soil and what were once green leaves will likely be forgotten.

Then there are the choices that resemble major limbs on a tree or perennials in a garden. Trees have many branches, but a limited number of major limbs, just as most people have about as many turning points in life as they can count on both hands. On the big limbs each leaf represents a subsequent choice that when we look back made a permanent, perennial difference in our lives. Generally, those enduring choices have a lasting effect on others also, such as the person you did or didn't marry, the education you did or didn't get, or the move to a far-away place.

In the same way that tiny leaves emerged on the oak's trunk after the little girl gave the bark her warmth, life-changing choices often begin as seemingly unimportant events. A friend phones and asks if you want to attend a party, you do, and an event happens that alters your life. Education, marriage, children, career choices, divorces, life-threatening disease, and the death of a loved one are all long range, and major turning points. They are forks in the road like forks on a tree's trunk.

The thousands upon thousands of leaves on a tree represent life's multitude of options and opportunities. The tree becomes its "choices," and so do you and I. Where a tree sprouts a limb will affect the swirls in the grain of the wood. Our major choices shape our lives as much as the way a tree grows affects the beauty of its wood grain.

Chapter 14

CELEBRATE

If you desire to know the goodness within yourself, plant a small fruit tree. See how the sapling's one or two little branches reach straight up to the sky. From the very start, the tiny tree quietly displays faith in an abundant, loving universe. That abundance is also there for each of us.

At night, while the city sleeps, you can stand by your tree and listen to the wind carrying the angel's whispers, encouraging the tree to grow. A whole chorus of angels is singing the same song to each of us.

As the months pass, notice how the tree makes little, yet positive changes. At other times the changes aren't even perceptible.

Outside a little cherry tree grows. It has three thin branches in a straight row, like a pitchfork. The sapling lacks roundness. It's still young and gangly. It could be pruned to force branching, but I decided to leave it alone, trusting that it could grow fine on its own. I did not want my opinion to determine what the tree would become. Whatever changes trees make, the changes are slow, but the progress steady. There is a great display of patience in the way a tree develops through the years, a reminder that heaven is very patient with our growth also.

As time passes, you can observe how each season your tree flows with nature's rhythm. It adapts to the seasons by adhering to an inner clock that does not stress if spring is early or late. If the groundhog stays in the ground, so be it. The tree serenely waits to open its buds rather than risk freezing its delicate petals.

The year begins for the fruit tree in the fall when children are ready to start back to school. Most children start kindergarten at age five, which is also when fruit trees are ready to bear a lot of fruit. By early autumn the tree fills its branches with barely noticeable buds that become the fruit for the following year's harvest. After the tiny green buds are in place on the branches, the fruit tree puts aside its summer garb and lets go of its leaves, much like we in colder climates put away summer clothes and get ready for winter. Throughout the wintry weather the little buds will stay warm in a protective brown covering. Sometimes snow will cover the branches further protecting the fragile buds against sub-zero temperatures, a reminder of how much nature provides for life to continue. Thousands of times we are lovingly protected too, although it is done so silently that we are often unaware.

Eventually spring showers will fall. Warm rain will seep deep into the ground, stimulating the sleeping roots. Later, when the tree's buds suddenly open we look outside and find the branches adorned in glorious blossoms. This is the time when it's fun to follow nature by getting dressed in spring colors then do something joyous.

When the temperatures soar on a hot summer day, you may possibly pick a piece of ripe fruit fresh from the tree. A broad smile might spread across your face after you taste the sweetness. To eat ripe fruit off a tree is one of life's simple pleasures. In that moment when you take a bite, you can savor some sweetness in your own life. With each mouth-watering nibble remember that the tree didn't let a bitter winter stop it from producing that piece of fruit. Nature shows us not to limit ourselves with negative thoughts because our thoughts and actions turn into reality as surely as the buds that were set a year ago became the fruit you pick off your tree.

In autumn the tree will drop its leaves. The majority of autumn leaves will likely stay in the yard to enrich the area closest to the tree. This is nature's way of reminding us that in life the greatest difference we make starts with those closest to us. A smile, a helping hand, a shared joke, a hug, a gentle word can do so much to make a difference in the environment within a family. Like the little leaves on a tree, it's the little things taken together that create a healthy atmosphere.

As you move about your neighborhood, you might observe how your tree is unique from other fruit trees of the same variety. Every tree is different, as is every person. When you celebrate how exceptional your tree is, also celebrate that there is no one else in the world quite like you. A tree has integrity to be only what it is. Perhaps this is nature's way of saying that anytime we imitate someone else, we end up in second place, a bogus presentation of our true self, and then our family and community suffer a loss of our full potential.

Sense your tree's inner magic. It has an incredible will to live through the storms that come and go. While the tree changes from year to year, you can observe that it protects itself from the bugs that infect it since it is far stronger inside than it appears. The tree is imbued with ways to heal itself the same as you can because God wanted us all to have those abilities.

If several years from now when the tree is large enough for you to lean against it, you might reflect on the wonder it holds. In that moment, celebrate because you are part of nature's goodness and grandeur and it is part of you. Through the years the tree gave many gifts just as you have done in your own life.

Young oak tree, Thousand Oaks, CA

Chapter 15

CAPTAIN JAKE

This is the story about celebrating a person, just as he is. The person I'm about to tell you about is in some ways very different from me, yet we get along very well together. Through him I continue to grow emotionally and spiritually because we grow within relationships with family, friends, and lovers the same way a tree grows within the soil that surrounds its roots.

Relationships are great teachers. Some are as permanent and solid as an oak. Some loves linger in one's memory, as much as standing deep in a grove of Giant Redwoods is an unforgettable experience. Occasionally, an attachment reminds us of prickly Saguaro cactus, which in the future warns us to avoid getting too close. Jake stirs images of a coconut palm because it is complex and unique among trees, as he is among men.

I think of Jake, first of all, as a man's man. I met him on a sun-baked July morning only hours after my mom died. Walking alone along the beach, my world seemed as unsteady as the sand under my feet. He came along and we started talking. He was as different from the high-powered businessmen I normally attracted as a palm tree is from a Christmas tree. I call him Captain Jake because he's the master of his own ship. At five-ten with Richard Gere hair, he has a twinkle in his sky blue eyes and a body made for 501's.

One of many reasons Jake reminds me of coconut palms is that he has always has had an affinity for water. In his Malibu surfer days he rode the waves up and down the California coast. At that time, he was a man drawn to tropical seas and women that inhabit steamy islands, or more correctly "women with whom he wouldn't want children." Committed relationships, putting down roots, and having children always topped his list of things to avoid in life. In contrast, I always knew children, a home where friends gathered, and a place in the community topped my list of desires.

Because Jake has avoided permanent roots, his similarity to coconut palms continues. Unlike oaks, which wrap their roots around deep rocks to stay anchored in one place, palms have a shallow compact root ball that sits in the ground much like the anklebone fits in the foot. The rounded anklebone allows a person to sway on the dance floor the way palm trees sway in the wind. Since coconut palms have shallow roots, they are easy to move. Landscapers need only dig a small hole to remove a huge coconut palm and haul it off to a new location. Lacking deep roots, Jake is equally ready to "press the button" then move on whenever and wherever opportunity knocks or the mood strikes him. He has no fear of dangerous situations, or even death, which increases his sense of freedom.

The coconut is similar since without warning it plummets to the ground. Then the coconut rolls down to a sandy beach and hitches a ride on the tide. When a coconut floats out to sea, it seems to shout, "I'll just do it and work out the consequences later." With equal suddenness, Jake joined the navy before he finished high school or told his parents he was leaving. They expected him to use his scholarship and unusual talents to study architecture as he had planned. It was seven years before they saw him. Yet every year on his birthday, wherever he was in the world, he sent his mother a dozen roses to thank her for giving him life.

Just as a coconut merrily bobs out of sight, once Jake decides its time to go, he lets nothing or nobody stand in his way, not even love. Even years later, his desire for new experiences still imparts a sense of bachelorhood into his closest relationships. He doesn't intend to hurt anyone by his exits, and he's sensitive enough to feel badly when he causes another pain. It's his eagerness to embrace change for the sake of change that stirs insecurity in others, and creates twinges of abandonment, for one never knows the day he'll leave, or if he'll return. That's the way he is.

I was the first woman that he wished he'd had children with and it left him with regrets. He, who had lived life on his own terms, never expected regrets or a sense of failure that he'd missed something important, and it upset him. When he left, the issue wasn't his leaving per se since it was expected as part of his nature. Besides, I, like a watery sea, did not want to decide the coconut's direction. Water understands freedom because it gets nasty, if it is forced to become stagnant in a container. It has always been nature's plan that water must flow and encourage things to grow. It's only when water is held back, dammed, ignored, that it rushes to find new directions, so it can flow gently again.

This brings me to another quality about the fruit on coconut palms. Before one gets to the white coconut meat commonly sold in grocery stores as shredded coconut, there are other important layers. The coconut that drops off a tree is about the size of a basketball. The outside is waterproof, smooth, tough as leather, and green—just as Jake was at seventeen when he left home.

The second layer consists of several inches of a thick fibrous substance much like a reed raft. The fibrous layer gives buoyancy, enabling the coconut to float alone on the ocean; Jake stays afloat too, drifting alone on numerous adventures. These two layers, the leathery and reed-like ones, are defensive and difficult to penetrate. Jake also possesses tough, hard, protective layers. In his blue eyes, one sees shrewdness, for he's not a man that's easily fooled, and certainly not someone anyone would want to cross. He's extremely tolerant and quite difficult to provoke, but his sting is numbing. He's capable of getting even with stealth and cunning, like a scorpion that whips and stings with its tail after it's behind you. I am thankful, I've only seen his sting, but never felt its force.

Now comes the third layer, the small dark brown coconut often sold in supermarket produce sections. Its rock-hard shell protects the delicious inner heart of the coconut. Under Jake's tough exterior, there's a side of him that's as sweet as coconut. He is essentially a very kind man. Fierce dogs become docile in his presence. Children instinctively trust him; perhaps they sense he'll be straight with them because he's true to himself. He is unique. When a recipe calls for coconut, there's no other food that can be substituted for coconut. He is that irreplaceable.

The last layer of the coconut contains the pure white meat and coconut milk. The milk sometimes is called the pure nectar of the gods. Within Jake there is a strong spiritual core. I can hear the awe he has for the universe when he talks about the constellations above, as if he's been to heaven and back a thousand times. It's his spiritual core that keeps him afloat in difficult times and rough seas.

Jake challenged me to understand unconditional love and its connection to freedom. As time passed, I learned through his departures that loss is an illusion. Affection doesn't end, it continues from a different direction, for love has many facets. It was in risking to re-examine previously held ideas of what love is that I was able, like water, to go where water never flowed before. A freer, more creative me emerged.

The greatest gift I learned from Jake is that just as there are many different trees, there are just as many different types of people. Each person is a certain way for a reason. I may not understand the reason to know how they fit in with the Divine Plan, yet each person is a treasure in the promise they hold to teach us about ourselves and further our growth. That is why I celebrate Jake.

Chapter 16

TEACHER OF LOVE

While writing this book, I tried to remember when I formed such a deep fondness for trees. I knew early in life that I loved trees and learned a lot from them. By the time I turned four my knowledge of trees became intimate. Jack, the boy next door, had a big old cherry tree in his backyard. When the cherries ripened, we practically lived in the cherry tree, which had the best cherries I've ever tasted in my whole life.

When I was in grade school, the dreariness of gray winter days disappeared when nature laid a puffy blanket of white snow over the landscape. Late at night, I'd quietly get out of bed, get dressed and tiptoe down the stairs to walk alone in the fresh snow. The giant elm tree on the corner sparkled under the lamplight. Its long hanging icicles turned into ribbons of diamonds under the streetlight's glow. On the branches, shimmering crystals of snow glistened like twinkling stars. I learned to look for miracles. To this day, the child in me awakens when snowflakes fall.

Perhaps my affinity for trees began in junior high while sitting on my girlfriend's porch talking about boys. From that porch, there was a view of a nearly flawless blue spruce that grew on an island in the middle of the cul-de-sac. Many times I admired the spruce when I went to her house.

However, the spruce suffered as a result of me getting a driver's license. I accidentally backed my Dad's red Mercury convertible into the spruce. The tank-like car survived, but my girlfriend was planning her wedding by the time the blue spruce filled in again. The fact the spruce returned to its former glory taught me that time does heal many wounds. That knowledge sustained me when I suffered a painful loss.

When I started loving trees is impossible to pinpoint, but more importantly I learned about loving on a subliminal level through a fascination with them.

The cherry, elm and spruce trees started growing before I was born. For the first ten years of my life, my parents didn't have any trees since they rented an upstairs apartment. Because we didn't personally own a tree, I came to sense you could love something without owning it. A person can't buy a tree from someone's yard anymore than real love can be bought or sold.

A few miles away from where we lived there was an apple orchard. In the orchard the air smelled sweet in spring when the trees blossomed. By summer the orchard's shade proved cool and soothing on hot, muggy days. In autumn, when the apples ripened, the cider from the local mill was a refreshing treat. Harvest time meant picking boxes of apples that found their way into my mother's pies and my lunch box. Everything about the apple orchard gave me a sense of abundance, and that abundance was an extension of God's love for us. When I ate anything made from the apples, I was sad to take the last bite. I still love apples. The whole orchard experience taught me that love delights and takes pleasure in the moment and in that moment knows joy.

The rhythmic sound of red apples hitting the ground signaled the final days of autumn as surely as canons at the last part of Tchaikovsky's 1812 Overture signaled the end of the fireworks display on the Fourth of July. All the trees in the orchard seemed like instruments playing in harmony to produce a bountiful harvest. In their performing together, I sensed teamwork strengthened rather than weakened the orchard. Teamwork meant that cooperation to produce something beneficial would not produce loneliness, destruction, or despair, but like the orchard togetherness, growth, and a bountiful harvest.

The fruit trees provide a gift that benefits our health, without the trees having expectations that we will care for them. Love is much the same. Human expectations of what another will do in return for our giving are like broken branches that become pegs on which to hang our disappointments. We can't require a person we love to love us back exactly the same way we love them without the potential to incur some discontent. We are each different. Another can only give what they have to give according to his or her understanding of love, and that may not be familiar to us or satisfying; in fact, it can hurt. Love and freedom let us select or reject according to our tastes. I love apples and men, but not all varieties.

While walking to grade school, I sometimes passed the time, trying to identify the neighborhood trees for a Brownie badge. Among the trees were red oaks, white birch, and a black walnut. In the fall, leaves of red, yellow, and brown fell to the ground. Wind mixed the leaves together. The leaves eventually became dirt same as humans return to dust. It seemed that no matter the color given different trees, in the end, one tree was no more or less than another, just as I was no more or no less than my classmates. Now, years later, after discovery of DNA, the Human Genome scientists confirm that people of the *same* race have more differences than there are differences between the races. We are essentially the same biologically and equal in the web of life.

There were several maple trees outside the front of the apartment that I could see from the mahogany desk, where I once danced before it became the place for doing homework. Because I tended to look out the window and daydream, I noticed subtle changes in the maple trees. As long as they kept changing, they weren't dying. Instead they were living, fully involved in their own adjustments and alterations. Everything the trees needed already existed within. I watched as each maple tree experimented with learning how to reach its potential. I came, after a long time, to feel that each of us also is born with what we need inside us to reach our purpose for being here.

Each week the maples slowly initiated changes with a new leaf here, a young twig there, until the twig and leaves were transformed into a branch. I loved the liberating sense of wonder that came from seeing trees grow through the years. It gave me the sense that we too can grow and even re-invent ourselves. I came to believe that, just as the maple tree is different from the cherry tree, each of us is imbued with the ability to experiment with creating our lives into something that's truly a reflection of our individuality.

As a little girl, I used to climb to the very top of Jack's cherry tree, high above the rooftop of the house next door. Touching each branch, feeling the tree, learning where to put my hands, where it's limbs were weak, and where the tree could easily support my weight. I'd wrap both arms around the main trunk in the same motion as a hug. My mother worried I'd fall. Sometimes she would saunter outside, walking far too casually, when I was at the top of the cherry tree. I'd see her lips moving, mulling over whether or not to tell me to get down. But, I never fell. I trusted that great rooted tree would stay there. It was an unspoken promise that the tree wouldn't suddenly break apart. So I learned from trees that when you love someone you have to be able to trust that they will be there as long as they can.

Chapter 17

NATURE'S ART

I returned to my car with my head down against the rain. It wasn't until I drove out of the parking lot that I noticed one of the most magnificent oak trees I'd ever seen. It stole my breath. The oak was a work of art.

Having studied art, I knew paintings tell volumes about an artist's "essence" because they reveal the truth within the artist at the time the artist created the work. For example, artist Mark Rothko's classic, meditative works suggest an eternal landscape. He painted long vertical canvases covered with stacked horizontal bands of bright pastels. After Rothko's wife died, his canvases lost their vibrancy. His later paintings displayed mainly stacked bands of depressing gray and black. His joy when his wife was alive and his despair after her death clearly showed itself in his paintings. Rather than give viewers clues to his thoughts by assigning word titles to his paintings, Rothko assigned numbers to his canvases. He believed that silence let his paintings speak their truth to the viewer.

Nature, like fine art, uses no words but speaks to us in silence. John Newton, who penned the lyrics to Amazing Grace, noted nature's unspoken attributes: "There is a signature of wisdom and power impressed on the works of

God, which evidently distinguishes them from the feeble imitations of men. Not only the splendor of the sun, but the glimmering light of the glowworm proclaim His glory."

I alone can't fathom the fastness of the Almighty, but many of us have glimpses of a higher power, and call God many different names. Although we don't know which name is most pleasing to Him, together with the universe, we represent the sum of the Divine.

The closest I came to understanding the Divine came in my early twenties during an illness when I had a near-death experience. I stood before the Great Spirit, which is the most accurate description I can give, because I couldn't see His face or body, only the lovely light. All conversation was telepathic, instantaneous, and silent, like nature. What struck me most, and I kept remarking about it to Him, was the mind-blowing intelligence and love that pulsed and emanated from God. I recall saying that I didn't know such love and wisdom was even possible. It was far beyond anything I'd ever comprehended. As we talked, I was aware of another quality. For the first time in my life, I felt totally free. When God asked if I wanted to stay, I had no doubts the choice was up to me. That part made me uncomfortable because I didn't want to offend God by leaving. Yet, I felt no hint of judgement, if I chose to return. I was sad to leave and to this day I can become teary-eyed with joy remembering, but I chose to return to my fiancée and the children we would have.

Looking back, I've often thought about those three intertwined qualities—wisdom, love and freedom—that were profound and unmistakable, existing individually and yet as inseparably as the three sides of a triangle. If one side of the triangle were removed, the triangle would cease to exist. I believe the three qualities are equally indivisible in God.

All that is natural reflects God as surely Rothko's paintings mirrored him. It would probably take me years to list all the ways I see the wisdom, love and freedom in nature, so for the sake of time—yours and mine—I'll only highlight two: spiders and dandelions. Both spiders and dandelions illustrate the saying that God is in the details.

A few weeks before my plum trees are covered with juicy ripe fruit dozens of spider webs appear between the branches and in the lawn. Any fruit I can't reach with a ladder I leave for the birds and other creatures. Shortly before the first plumbs are ripe enough to drop from the tree, a huge convention of assorted insects arrive. I appreciate the fact spiders appear at the right time and place. Then they kindly move on to my neighbor's since his fruit tree ripens after mine.

The spiders are part of nature's ingenious checks and balances. The spiders curtail the insect population by building webs to keep nature in balance, since the insects proliferate due to the abundant food supply of ripened fruit.

The insects eat the fruit and in return leave deposits that quickly break down the fallen fruit so garbage doesn't pile up. Within a week or so there's no more traces of the fruit that has decomposed to nourish the soil.

Sometimes I study the circular spider webs, finding in their exquisite geometry mini-works of art. Spider silk is finer than human hair, lighter than a little down feather, and stronger than steel. Only recently are scientists even remotely close to duplicating ingenious spider threads. The scientists are putting spider genes into cows and alfalfa fields so they can extract the protein that makes the silk. It's a lot of work and expense to do what a spider does in relatively little time at all.

Spiders that weave circular webs that contain several different kinds of silk that are brilliant in their engineering. The strongest spider silk is so strong it allows the insect to fall from great heights. Other strands are sticky enough to catch insects while spiders make a third type that allows a spider to go to the center of the snare without getting caught in its own web.

Birds fly into the plum tree all the time to eat the fruit, but are unlikely to get their wings caught in a gummy web. Spiders, which haven't changed in over 125 million years, weave webs lovingly engineered to reflect sunlight in a way that the eyes of birds can see the web from afar.

Nothing natural is irrelevant. As a case in point, consider the much-maligned dandelion. In early spring, if you touch a dandelion, you can feel the heat it generates in comparison to the lawn's coolness. The bright yellow blossoms are like a bit of sunshine that warms the air to protect fragile spring flowers, such as crocus, on chilly spring mornings. The dreaded dandelion does so much more. The blossoms, when simmered for a few minutes then placed on the skin, are said to help clear acne and discourage skin tags. The leaves help cleanse the liver and serve as a diuretic when bloated. In August, when it's sweltering, dandelion greens help lower the body's internal thermostat. The next time you add dandelion greens to a salad remember that nothing about you is irrelevant. When the dandelion blossoms are spent, its seeds travel on the wind, carrying an unspoken promise that life continues. That promise is in every spring that follows winter, reminding us that death is not final. We are loved too much to become obsolete.

An artist with words, Anne Frank, understood what was behind the splendor and mystery in nature. She wrote, "The best remedy for those who are afraid, lonely, or unhappy is to go outside, somewhere where they can be quiet, alone with the heavens, nature and God. Because only then does one feel that all is as it should be and that God wishes to see people happy, amidst the simple beauty of nature."

Nature is silent, but speaks to us in a language beyond words, a language we learned before we could talk. The language came to us in infancy in a slower timeless stage of life. As babies we did not know the words, wisdom, love, and freedom, but in our innocence we sensed them and took delight in those qualities, and found joy.

Oaks along Malibu Creek, Malibu Canyon, CA

Nature's unspoken language provided the freedom needed for our own experiences in nature to become more personal. By the time we were old enough to play on the grass, we smelled the earth's slightly woodsy, spicy scent as we moved across the lawn on our hands and knees. When we scooted to the flowers, our uncluttered brain registered the blossom's sweet perfume, and evermore we associated flowers with love and beauty. Free to move within a given space, we went to a tree and put our hands on its uneven bark, then using the tree for support we stood with the tree's help. In the back of the mind, we would remember that trees support us. The first time we felt the bark it felt strange. The nerves of our fingers sent the tactile impression of the bark's rich texture into recorded memory. Those sensory impressions prompted us to treasure the adventure of discovery and honor the wonder of life.

When we grew a little older, many of us learned more about love not only from our family, but also from one of nature's teachers, the family pet. The family dog or cat welcomed us home from school, understood us even in silence, romped freely at our side, and lovingly kept watch over us while we slept at night. Heaven operates the same way.

The earth is the all-time best of gifts, a place where all needs were anticipated, not only for us, but also for the smallest insect. Food, water, and the means to obtain shelter were lovingly, wisely and freely supplied to every creature. The earth's support systems were not put here for the Almighty to get rich, but to support the life that is here on the earth. Although humankind was given curator status, taking care of all that lives upon the earth was not designed as a forbidding task. Our work was made easier because in an unselfish blueprint everything in nature enhances and assists something else.

The sun and rain encourage trees and crops to grow, and in turn they provide food that strengthens the creature and critters that need to eat. In case the free wind doesn't scatter seeds to keep nutritional and medicinal plants flourishing, birds act as back up to take seeds and enrich another area. Leaves from trees replenish the soil in temperate climates. Oceans send moisture into the clouds, which the wind takes around the globe without stinginess so that every continent thrives with life-giving rain. In meadows, flowers of every hue and variety amplify life's beauty and sweetness while fruit and nut trees are things that affirm God's generosity. All the above comes with maid service in the form of ingenious, biodegradable recycling.

Intelligence, love, and freedom are innate in us and in nature because the artist, the Creator, put them there. The artist can only create a representation of his or her inner self. God loved us enough not to wish to control us. As Jonathon Livingston Seagull said, "If you love something set it free, if it comes back to you it is yours," and so we were set free so that we could return, not by force, but voluntarily to the Almighty. Anything less than our returning because we wanted to would diminish a God of love, freedom and wisdom.

Around the world, people prize fine art and collectors pay millions for particular pieces. In honoring artists living and dead, we seek to protect what they created as a celebration of human creativity and divine inspiration. Then we build museums to house their art to share their work with present and future generations.

In this and coming centuries, I believe, we will give more tribute to the most remarkable artist of all time. We will become more careful curators of the Creator's art in the living museum we know as Earth. When we do, we will have evolved to a widespread understanding that all the beauty on earth is also in us.

Chapter 18

THE AMAZING SEED

This is a story about an amazing seed. Since its discovery the seed has grown into a tree with enormous proportions. The seed represented an idea. The idea was given a name, Evolution. The seed grew because in it there was truth. The tree kept growing until its branches spread from one side of the country to another. People on other continents heard about the amazing seed. Over time the seed mutated until its deeper significance became forgotten.

The amazing seed started as a tiny seedling in the midst of a deeply established royal forest that had thrived for thousands of years. Life in the forest appeared to change little over the millenniums. Horses still looked like the horses in ancient cave drawings made by ancestors eons earlier. Indeed, people agreed the horse was unchanged. Furthermore, no one living could recall when trees in the forests had changed in a lifetime, nor had the dragonfly or turtle. The wise men told the people that everything that God made remained the way it was created, or it perished. The people became afraid of change.

There was a book in the kingdom that gave testament to a time long, long ago. Part of the Good Book told how God had appointed Moses to lead a nation across a desert and out of slavery so that they could be free. Instead of understanding that freedom is part of every soul, something else happened.

In various parts of the world, leaders heard the story of Moses and equated it to their own life. With pomp and ceremony, leader after leader told the people that God had chosen him *to rule over them with a firm, guiding hand.* The people believed what their leader told them and accepted the king as God's representative on earth. They became subjects, which meant they were *subject to the monarch's will.* And so it was that royalty, in varying degrees, assumed the power to pardon or punish anyone, to raise up or cast down low, to give life or take it, and to judge, but not be judged by any person other than God.

As generation after generation accepted their leaders' authority, monarchs grew ever more powerful along with their armies. For thousands of years—from the time of Egyptian pharaohs to Mesopotamian kings, to Chinese emperors, to the Aztec emperors, to the Roman caesars, to the Russian Czars, and to Europe's medieval kings—monarchs had power over those they ruled. It proved an ominous day when ego ruled. It furthered dependency on the aristocracy in particular, since the ordinary person held little or less power. The domination of society by hereditary aristocracy thrived alongside ignorance, superstition, and tyranny.

The idea of monarchs, who were part God, persisted because it also contained truth. There truly was a spark of the Divine in those who sat on thrones. But, what their royal subjects didn't experience was that there was an equal and identical spark of the Divine in them and in all that flies, crawls, swims, and walks upon the earth. Because the ordinary people—the serfs, peasants, commoners, or Lessers as the Betters sometimes called them—were made to feel unworthy of great privilege, many believed it. Much unhappiness ensued. Dreams went unrealized. Since the majority accepted their lot in life as unchangeable, it enabled hereditary titles to remain intact for centuries. The favored place of the aristocracy allowed them to preserve their privileges in business and society.

With hereditary titles being passed down, the rights of the aristocracy became as fixed as the rights of the monarch. The rigid, unchanging hierarchical structure mimicked the idea that nature didn't change, but stayed the way God wanted it. Monarchs and the aristocracy meted out harsh punishment to those who dared not conform to their authority.

Monarchs received help in furthering their position from state churches. Like a set of scales that seek a balance point, churches sought to balance the great power of monarchy as an institution by becoming equally powerful. If a church leader or monarch met with either institution's disapproval, it incurred the other's mighty wrath. To keep both institutions' power intact, God often was cast as an angry, punishing God—who like society's "Betters"—could extract terrible reckoning for not following God's commands. In many churches, sin spurred lasting shame that reached into the minds of people while thoughts of eternal damnation created fear. So it came to be that shame and fear lived side by side like the right and left hand on the body.

After thousands of years of rule, monarchs around the world became aware of the amazing seed, Evolution. It happened when Charles Darwin put what he knew about the amazing seed into a book titled *The Origin of the Species by Natural Selection, or the Preservation of Favored Races in the Struggle for Life.*

The idea that nature changed, and didn't stay the way it was created, shook the world.

If a species changed, then might it imply that the monarchy would also? If the monarchy changed it would alter who sat on the throne and wielded power over others. Or, worse still for the aristocracy, no monarch would control the country as happened in that upstart nation across the

Atlantic. Approximately fifty years before Darwin took his famous voyage in 1830, the American then French Revolutions took place. Replacing the monarchy with democracy was a horrible thought for some. The buzzwords that became an affront to thousands of years of tradition were adaptation, change, mutability, and impermanence.

* * *

Across the globe, monarchs and their counselors gathered to deliberate whether to let the "amazing new seed" grow or promptly exterminate it, for they had that power.

In a council chamber, a monarch's advisor said, "Your Highness, in Darwin's title I belief there's a loophole."

"Speak up." From his throne, the monarch eagerly leaned forward to hear what the man had to say.

Adjusting his glasses the advisor said, "It's in the phrase '*Preservation of Favored Races.*' Royalty is indeed favored, is it not, Your Majesty?"

"Ah yes," the monarch agreed, gazing around the gilded throne room.

"In centuries past, your Highness, usurpers to the throne attempted to grab control of the government, but those that engaged in treason failed."

One of the monarch's advisors narrowed his eyes then a grin broke on his face. "Good point," the counselor said. "All of the usurpers' failures validate that only the fittest survive. Our monarch's line is the fittest and therefore is entitled to sit on the throne."

One of the landed gentry rose and spoke for the other nobles. "And what becomes of us, your Highness?"

The monarch pondered the question. "The titled of the realm also validate survival of the fittest."

One of the wise men cleared his throat. "Darwin's work affirms that class divisions and racism are not only acceptable, but also scientifically respectable."

"Are you certain?" the monarch inquired.

"Yes, your Majesty. I've gone over his papers several times. Why, even the author's first cousin, an equally brilliant scientist, Sir Charles Galton, is working on a system he calls eugenics that supports your favored position."

"Go on," the monarch ordered for he had heard of Galton's work, but not the details.

"Eugenics is devoted to improving the human species through control of hereditary factors in mating." The wise man stopped to glance at his notes. "He advocates controlling births through sterilization of deviants, criminals and those with lesser intelligence in order to elevate the race."

While the monarch's council deliberated, in another part of the kingdom, church leaders of different denominations gathered in their respective churches for they too knew that the amazing idea had serious implications. One of the bishops summed up the sentiments of those gathered. "The Bible says the earth was created in six days. If Darwin is right, people will think the Bible is wrong. We can't have that."

A young man scholar said, "Why can't both be right?"

"No," the assembled intoned. "Every word is the absolute truth. What you're suggesting is heresy."

"Why must we live in a rigid world of black and white? Why do we create these separations?"

They were about to condemn him, when an elder came to his defense. "God Himself named the time span 'a day,' but no human was on earth during earth's formation to know how long 'a day' was by His heavenly standards."

They went off to deliberate. When they came back they said a day is twenty-four hours. This Darwin chap who says he can prove birds' beaks change wants to change too much.

In another part of the kingdom, scientists of the Royal Academy of Science gathered together. They came because everyday the amazing seed was growing and producing new ideas. Zoologists, botanists, philosophers and a host of other scientists had each found evidence to support the truth of the amazing seed.

The scientists deliberated. They strolled outside past neat boxwood hedges bordered with primroses. One of the scientists remarked, "I admire the fine sense of order in aristocratic gardens. There is nary a weed or a spent flower out of place. Such perfection delights the eye."

"Indeed, it does." His associate surveyed the vast expanse of rolling green lawn. "The monarch would no more abandon or neglect his lands to nature's wiles than abdicate."

Upon entering a yew maze, they marveled at its size, and the firm guiding hand needed to keep it tidy. The older man remarked, "What a terrible loss if the precise arrangements of our Victorian world were disrupted by evolution being carried too far."

"Are you suggesting chaos would rein?"

The senior scientist nodded.

"But, Sir, order and disorder seem to always dwell together. The gentle rain can become the fierce storm." He pointed to a bed of larkspur saying, "The flowers are ingeniously constructed with such care, but the flowers throw their seeds to the wind."

Wishing to elaborate for his younger companion, the scientist suggested, "Look around at the gardens. From every vista you see the monarch's *firm guiding hand,* governing the same way as God governs." With the lift of his eyebrow, he added, "Darwin says that Creation is *unguided and without purpose.*"

"Much like we consider a vacant lot, if it's left to nature and not made productive, we say it is abandoned."

The younger scientist pondered a bit. "Sir, then science has to consider that *God abandoned and neglected humanity at the dawn of creation by withdrawing His firm guiding hand with natural selection.*"

The conclusions of science became so pervasive that it led eventually to the notion that God was dead.

In a century and a half since people probed the possibility that God had abandoned us, much evolution in human thinking has taken place upon the earth. With the Civil War slavery ended, education became more widespread, and the middle class grew dramatically. Many monarchies in the Old World gradually tumbled. Numerous inventions—electricity, cars, planes, movies, telephones, television, computers and satellites—brought diverse people together to share ideas. As the debate over creationism versus evolution continued science found huge gaps in the evolution of some species. The cockroach, dragonfly, crocodile and turtle had changed little since the dinosaurs. As scientists tried to understand why change was unpredictable, the knowledge that the universe could not turn out so well by accident alone became clearer.

There are still leaders that believe they have a right to have power over others. They do not understand a Divine love that would endow freedom to all, even an insect. However, the connection between freedom and love was clear to people who had escaped from living under a dictator, or who understood the dynamics of abusive relationships. They knew that love is kind. It's not controlling, intimidating, threatening or violent. It does not instill fear or shame.

The amazing seed that provides freedom for all on earth to evolve, if desired or necessary for well being, has been here since the dawn of time. Freedom is an amazing gift. It is inalienable, a right that can not be taken away or separated from us. In the darkest of times and situations, we retain by the "Laws of Nature and Nature's God . . . the right to life, liberty, and the pursuit of happiness."

Chapter 19

CREATIVITY

In Southern California, there was a year with a spectacular spring. Everything seemed to bloom at once and then it turned cool enough so that spring lingered for weeks. The creative potential in nature came alive through a brightly painted landscape. Purple jacaranda trees bloomed along with yellow daffodils. Bougainvillea in bright magenta spilled over wrought iron fences. Flowerbeds burst with colorful red and white striped petunias. I wanted to pinch myself to make sure I wasn't dreaming. Never had I seen so many plants in bloom at once.

Color greatly affects me. I hear music when I see a meadow in bloom. Sometimes I get an urge to sing like an opera singer in the multi-hued fruit and vegetable section of a supermarket, but I usually restrain myself.

I'm filled with wonder at the beauty in this ever-evolving, ever-changing and constantly creative earth. The idea itself fills me with no small amount of gratitude. Compared to other planets, if you love color, earth is definitely the place to live.

The planet reminds me that God is a creative genius who put a bit of His creativity in everything. Earth was faultlessly designed to change from minute to minute, day to day. The universe constantly, lovingly mixes colors for us.

It begins at dawn then continues for twenty-four hours without commercial interruption. As the light comes up it reveals red raspberries on a vine, wild poppies waving in the wind and the sun dancing on the blue sea. Then night brings the evening star show, silver moon shadows on the lake and misty fog among the trees.

I remember when I was a little girl in Sunday school, the teacher said, "Everyone is made in the Creator's image." I looked up at the painting on the wall where God's picture hung in a gold frame. He wore a long white beard and sat on a red velvet throne with carved gold trim. As I sat on the folding metal chair wearing a ruffled dress with polished leather shoes, I didn't like what the Sunday school teacher said about "image and likeness." I wanted to get up and find my parents to ask them if that was true because I feared I'd wake up one day with a beard and turn into a man. I knew even then that I liked being a girl.

When I grew older, I began to suspect that "image and likeness" meant far different than I surmised in childhood. In art classes, I realized it's the creativity within us that connects us to the Creator. Perhaps that's why the creative process usually makes people feel joy and peace inside. Creativity is healing. It begins by quieting the mind and in the meditative process we touch our spirit and reach inside ourselves to find the truth within. Then inspiration comes along with the energy to make the vision a reality. Once creativity becomes as integral as breathing, then that creativity applies to other areas of study, life, and relationships.

The great artist Pablo Picasso rejected attempts to explain his art in technical and intellectual language. He said, "We should be able to say that such and such a painting is as it is, with its capacity for strength, because it is touched by God."

The greatness of the universe goes far beyond observing its grandeur and fabulous colors. The Almighty's most extraordinary work was in creating a world without end in which change is part of everyday so it is never boring. A world that didn't need to be controlled, perhaps to free God so he could go on creating elsewhere because a creative person lives to create. Earth isn't done, nor is the universe; it was designed with the freedom to go on re-creating itself. All is recycled and made new. The mountains turn to sand. Volcanoes make new mountains.

The same creative force is imbued in each of us. We create our world and our reality with every thought. We create constantly, be it a building, a garden, children, humor, toys, letters, or the family dinner. We have the choice to create as nature does, in a way that lends beauty and harmony for others in the world around us to enjoy.

Chapter 20

LIVING IN THE MOMENT

This is the story of how nature reminded me of the importance of living every moment.

Usually, gentle breezes off the Pacific Ocean blow on shore then make their way all across the United States then on to Europe. However, in late autumn for a few days the wind flow can totally shift with the arrival of the often dreaded, devil winds, known as the Santa Anas. The fierce Santa Anas come from the East where the Mojave Desert and Death Valley suck the moisture out of the air and heat it. The resulting dry, hot winds gusting at sixty to eighty miles per hour often fan the infamous brush fires that plague Southern California.

If the Santa Ana winds are mild and bring no fires, Los Angelenos breathe a sigh of relief and turn their attention to a celebration of Southern California's beauty. Because the winds clear out the smog, the Santa Ana winds create the bluest skies. The Channel Islands some twenty miles out in the Pacific Ocean are visible from inland mountainsides. At sunset, as the sun starts its descent into the sea, the mountains literally take on a purple-colored majesty. When darkness falls, the city's lights sparkle as if mile after mile of precious jewels lay on the valley floors.

It was on such a day that I went for a walk to feel the free wind. The Santa Ana winds were sending tumbleweeds rolling down the street like giant beach balls.

At the oak tree where the crows landed, I faced the trunk and I rested my elbows on the lowest branches as if leaning on a neighbor's fence. Standing there, I felt the limbs of the oak moving with the wind. Then, pretending I was a musical conductor, I stretched my hands out to my sides and placed the underside of my arms on the underside of the limbs. The branches then became extensions of my arms. Holding an imaginary baton, my arms moved in unison with the huge oak, as the tree became like a full orchestra playing a march in four-four-time.

The branches, as thick as a wrestler's biceps, grazed the top of my shoulders. As the gusts blew, I felt the branches raise and lower with the wind. When the heavy limbs pushed down on my shoulders, I was forced to exhale. When the rush of wind past, the limbs lifted, I inhaled. The closer I edged my body to the tree, the greater the need for long deep breaths in order to stay in sync with the weight pressing on my shoulders. As the branches moved up and down, I inhaled to the count of four, held the breath to the same count then slowly exhaled to the count of four. As I breathed in union with the balancing of the tree's trunk and branches, I became invigorated. The breaths were so deep I felt them in the depths of my lungs. The vital energy of the wind was circulating not just outside, but inside me, and I felt myself breathe away some earlier tension. The experience reminded me of the benefits of yoga breathing exercises.

Now and then, the four-count rhythm of the branches moving up and down on my shoulders changed. The first time it happened, I thought, "Hey, wait a minute! You're playing a different song. What's going on?" Moments later a howling gust of wind came roaring out of the canyon. The oak anticipated a new tempo before I could hear or sense the change. The branches already were swaying to a faster beat when the strong blast hit seconds later. Before this experience, I assumed the wind moved the trees. When the wind is in the trees, the tree isn't passive at all. It's reacting, responding, participating in what is happening. By its response, the oak truly lived in the moment.

I thought of animals that foretell earthquakes. I didn't think that plants might also warn us of changes. When I told Captain Jake about the experience, he said that the four-count was often the rhythm of the ocean far out at sea. He added, "It's the earth's normal heartbeat. It's very healing for us."

Now months later, it seems totally naive that I was surprised at the way trees work in harmony with the wind. The whole earth and all that lives here and beyond are interrelated. Ah yes, Oneness.

Chapter 21

THE FICUS VINE

If you stand on almost any California hillside, you can tell where water flows underground by where oaks grow. Plants have an innate sense about the best place for them to put down roots. They also know what to do, when to shed leaves, when to bloom, how best to grow, a fact that reminds me of an experience with a ficus vine.

Several years ago, I planted six one-gallon ficus vines along one side of a fifty-foot-long stucco wall. Five of the vines grew profusely. One didn't grow at all.

There's a saying that if you agree to accept only the best life offers, the best is what you get. I thought perhaps the ficus that didn't grow was infected with some kind of insect, but it wasn't. In time, I was reminded that plants, like people, are imbued with an innate sense of what is best for them. I was required to honor this rule.

The one ficus vine that didn't grow left a gaping hole along the wall. The reluctant ficus just sat in the ground like a puny piece of string. No amount of coaxing could cause the ficus to catch up to the other vines. I tried everything I could think of. I tickled the soil, aerating it at the base of the ficus. I fed the ficus the best fertilizer, and trimmed trees to give it more sun. Then I installed a drip irrigation system. Finally, I tried sweet talk. Nothing.

The ficus curled its branches tightly into the main stem so it appeared to tarry in an arms-folded frump. The plant wasn't angry. It was "thinking," but I didn't know that. During the next five years, it sat there not growing in any way notable. I spent so much time on it I didn't have the heart to yank it out, even though a neighbor recommended I do just that.

"You should get another one," Molly suggested, "and fill the hole in the wall."

"I'll give it another year," I replied, for some reason clinging to hope.

Another neighbor, Ms. Kitty, joined us. "Perhaps, it's staked the wrong way."

"What do you mean?" I inquired eager to hear from the voice of experience.

Ms. Kitty ran her arthritic hand through her snow-white hair. "Some ficus vines grow branches that curve to the right, other ficus curve to the left, just as some people are right-handed while others are left-handed."

"I didn't know that. Thank you."

She walked over to the bare section on the wall, picked up a stick and poked around in the dirt. "Yep, it's staked to grow to the right, but it's a lefty."

"How can you tell?" I asked.

"See how the three little branches curve counter-clockwise. You got it tied to grommets on the stucco in a direction contrary to its nature."

"So that's why it didn't grow."

She smiled kindly, "It's been trying to tell you, it's a lefty, not a righty. The poor ficus can't do it your way."

I nodded. "That makes sense."

"If that ficus yielded to the way you staked it then it would remain scrawny. You'd have eventually yanked it out of the ground and cast it aside." She bent over to untie the twist ties that held the ficus. Miss Kitty looked at me over her bifocals. "If living things don't follow their true nature, it only causes more sorrow in the end."

Her kind blue eyes twinkled when she found a little leaf emerging near the base. "This little plant's been busy. It only appeared to sit around doing nothing, but it's got a healthy dose of determination."

She pointed out that under the mulch the ficus was growing new shoots *below* the point where it was tied to a stake. The ficus reached down inside *itself*, below all the former layers of the past, down to its inner core, in order to begin growing anew. The puny ficus gained gusto after it freed itself of the limitations created by being staked against its nature. Within six months the ficus vine filled in the wall, growing over twenty feet during the summer. It caught up to the other vines that had been growing for several years. The ficus vine was spectacular when it followed its *bliss.*

The ficus showed that, if we've been a slow starter, or were held back in some way, it's never too late to recreate ourselves. That's what trees do in spring. No matter how much we feel weighed down, or how many layers of stuff we have to get through to reach our true self or our lifelong dream, take it a step at a time. The ficus spent years in preparation, which is a reminder that we are all works in progress.

Filled with trust and enthusiasm, the ficus gave itself permission to claim what it was really was. The little left-spiraling ficus accepted only being true to itself so that it could become the best it had to offer. When, like the mighty ficus, we reach for the best that is within us, we walk in truth and our spirit soars.

Chapter 22

THE ESPALIER

On the way to the oak grove, I pass a home where a man conscientiously trains three evergreen pear trees to grow on an elaborate lattice that spans the length of his garage. The man tightly draws his lips into a grimace as he maneuvers tender twigs into place. The young stems yield to his guidance the way children do when they're young and supple. The homeowner is giving the espalier a solid start in the direction he finds pleasing. He diligently keeps careful watch over the espalier's growth. Sometimes he angrily pinches off an offending leaf with a quick snap of his wrist. He makes sure the pear trees conform to the design that consists of large and small evenly spaced rectangles.

The espalier is twenty-two feet wide by eight-feet tall with steel foundation supports firmly anchored into the framing and stucco of the garage. Only an 8.1 earthquake might bring it down. Baring a quake, with time the espalier promises to become the most spectacular aspect of the man's garden, an object for friends to admire.

The espalier bothers me. It must play the gardener's game. In return, the espalier doesn't have to take responsibility for itself the way an oak tree growing beside the road must grow strong and become independent to survive. The gardener provides for the espalier's needs in exchange for conformity.

Oaks in the wild are by nature free and independent. They can grow in many directions—up, out, down, and over a road or stream—but with the espalier, there's no chance for such freedom. Any defiance by so much as a leaf out of place upsets the pattern. The system isn't set up for win-win. It believes in winners and losers. The espalier will be rewarded if it does what the owner has determined the pear tree is to do.

I have to admit I also had an evergreen pear espalier tied to a trellis in my backyard that surrounded a fence around the pool equipment. Green tape bound the branches to the lattice. From the moment the espalier was installed, I felt a bit despotic and that made me uncomfortable. Perhaps the espalier on the gentleman's garage wall is happy. My espalier wasn't. It constantly grew opposite to what it was supposed to do. It was a rebel, refusing to adhere to the pattern laid out for it on the trellis. I always felt I was preventing it from tapping into its natural beauty. I was messing with the wonderful, essential wildness in nature by controlling its future because heavily pruned evergreen pears don't flower. Not flowering is like living without ever reaching one's potential.

Before long, I ripped out the trellis, the way the wall was ripped out in Berlin. When my children asked what I was doing, I said, "I want the pear tree to be free to become its dreams." They smiled, knowing I was talking about more than the evergreen pear. "In return," I added, "I want it to give me the same respect."

You may not believe this, because I didn't either at first, but after five years I began to take notice of another change in the pear tree. The freed espalier dropped its leaves neatly at its feet where they stayed, and never again did more than one or two of its leaves end up in the pool that was only five feet away.

Each spring, it flowered magnificently with fragrant white blossoms. I was glad it was free to follow its highest vision for itself. It could never have flowered profusely, if I'd kept it tied to the lattice or clipped its buds to keep it compact and well ordered.

From the beauty of the evergreen pear and the oaks growing free, I'm reminded how much nature imbued living things with freedom, even at a cellular level. Cells know, as an example, when to fight disease, to grow and to take in nourishment. Every living thing deserves to be free to master itself.

CHAPTER 23

SING A HAPPY SONG

During the late 1960's and 1970's when "flower power" took on new meaning, the discovery that pots of African violets blooming on a windowsill actually responded to music became big news. There was even a study in which houseplants were hooked up to electrodes, and the plants responded to the cries of lobsters as they were plunged into boiling water.

About this same time, Dorothy Retallack at Colorado Woman's College in Denver conducted experiments on plants. She discovered that continuously played music "exhausted" plants, while a few hours proved refreshing. She also showed that plants die when only one note is played, similar to the ghostly monotone on a heart monitor when the heart stops. Discordant music associated with eliciting negative emotions, retarded plant growth and sometimes even caused a plant to die. Soothing, re-enforcing, uplifting music, such as chants, soft jazz, soft pop, and some classical, caused plants to grow and flourish.

As a result of the various studies, people started playing music and actually sang to their plants. Although some thought it weird to do so, not everyone was a skeptic. People that previously couldn't keep plastic flowers alive found Green Thumbs just by placing potted plants beside the stereo or on top of the TV. With this knowledge houseplant sales soared dramatically.

All this implied to me that there's benefit in amiable sounds, which may be why birds sing so pleasantly. Nature offers an ingenious symphony of gentle, flowing, and subtle sounds. Birds, whales, water tumbling over rocks, the wind in the trees, and other nature sounds are the concert heaven designed to keep earth healthy. Nature is full of unknown secrets about the way all things connect.

One connection that's very old is the link between birds and trees. Birds sing in trees, build their nests in them, feed on the insects, and are nourished by the fruit trees produce. In some great cosmic plan, trees and birds seem made for each other.

Learning from science that plants "hear" makes me wonder what trees hear. Trees wrap their roots around rocks in the ground. The crystals in rocks might transmit notes the way crystals were used in radio transmissions. If elephants can communicate by making and receiving vibrations through the earth, then why can't crystals in rocks transmit to trees the notes birds sing?

The other night I had a funny dream about an oak "listening" to two robins conversing on its branches. The dream went like this.

"When did you get back from Mexico?" the smaller robin asked.

"Just flew into San Francisco a couple days ago. Nice flight. Caught a thermal. We were really cruising."

The other one sings, "On our way here, we stopped at the Grand Canyon."

The big robin bobbed its head in understanding. "The Canyon's one of my favorite places, especially at dawn. Ever been to Disneyland?"

"No, not yet."

"Pretty flowers. Great place. You ought to stop there some time."

The oak continued listening with no hope of ever seeing such sights, being rooted as it was.

The largest robin asked, "If you've only recently arrived in the Bay Area, you must've left a few weeks after us. We took off when the whales called."

"You can't miss their low, unforgettable melody."

The bigger robin flew down and pulled a worm from the grass. After eating it remarked, "It feels like there's a storm coming. This El Niño is changing the weather. Did you encounter any storms?"

"Yeah, it was pouring in San Diego."

"I heard it's dry in the Pacific Northwest and Alaska."

"I heard that too." The robin flew back into the tree. "After the storm ended, we followed a luxury liner heading to the Bay Area. You wouldn't believe the food they throw off those big ships. We ate like pigs."

"I know what you mean. Well, gotta fly. Nice seeing you. It's spring again. I'm meeting my mate in Sausalito."

"I have to get going too. We're nesting over at Klamath Lake."

"Oh, where the bald eagles nest."

"Right."

"Have fun."

If it sounds strange to dream that trees can tune-in to eavesdrop on birds, here's a bit of mind boggling information that's not from a dream.

In 1994, the brilliant French physicist and musical wizard Joel Sternheimer figured out how plants respond to sound and conducted controlled scientific tests to prove it. By using a precise series of notes for a specific plant, he showed he could make a plant grow or not grow. Fundamental to these experiments was the role of proteins. A protein is a group of amino acids joined together in a series like a pearl necklace. There are at present twenty-two known amino acids, and they are the key building blocks of all life.

The French physicist combined science and music and

made plants grow by using very carefully selected notes to target a precise, miniscule amino acid in the protein of a plant. And get this, when the whole series of notes in the "plant's song" was played, the whole protein molecule, all twenty-two different amino acids, responded to "its notes." In other words, the right combination of notes electromagnetically "speaks to" that plant's internal cell structure.

Big deal, some may say. However, if this knowledge was implemented think of the applications. If you don't want the grass to grow while you are away on vacation, you could play specific notes and skip mowing the lawn for a couple weeks. You could slow the growth of weeds. You could increase crop yields. You could decrease the use of fertilizers that feed the algae and eats up the oxygen in wetlands, thereby helping to save songbirds.

If the amino acids in plants are genetically keyed to certain notes, there's surely something inside each of us that draws us to particular rhythms, keys, sequences of notes, and instruments. Some note combinations would have positive or negative effect depending on the individual. I've often wondered why since childhood I've responded more deeply, passionately to a piano being played than any other instrument. From Mozart to Gershwin, the Beatles to Billy Joel to many other musicians too numerous to name, when I hear their piano music, I'm immediately transported to a less stressful place.

With our bodies as receptors for sound, reverberating, responding at a cellular level to each note and chord, music becomes as important as food for the body. In turn, like the birds' songs that helps plants grow, the music we create resonates out into the world, affecting and shaping it. Music has great power to sooth, heal, lift, and bring us into harmony with nature's symphony. Its enormous healing power could become part of medicine in the future. Could music even bring peace instead of discord? Could heated arguments be brought down several notches with soothing music?

I've heard that each of us has a "happy" little song inside us, which we sang as toddlers. I re-discovered mine when I thought my mom was sleeping in her hospital bed. She opened her eyes and said, "You sang that as a child. I haven't heard you sing it in such a very long time." It was only six notes. If you don't remember your personal song, maybe someone in your family does. Whatever your song is, try and remember it. Then sing, sing your song.

Chapter 24

CALL THE DOCTOR

The oak tree that held the flock of crows was in trouble, and I knew it. My son suggested I call a tree doctor, so I got out the Yellow Pages and phoned a specialist. I told the tree doctor, "About five feet off the ground where the tree forks, the bark is dripping wet."

"What kind of tree is it?" he asked. After I answered, he thought a moment then said, "Oaks do that sometimes. It's probably nothing to worry about. The tree usually corrects itself."

Summer arrived in its entire splendor and dried the wetness. Birds sang melodiously in the oak. I chided myself that I worried too much for I had to admit that the tree's foliage looked terrific.

When school started in the fall, the moisture returned. Ants saw fit to crawl in two busy columns, one going up the trunk, the other down. The ants raced along the trunk like cars traveling along the highway. Their destination was the wet spot. I wondered why the ants were interested.

Around Halloween, Santa Ana winds blew at gale force. As I approached the tree, I stared in shock. An enormous limb, the length of a semi-truck, was lying on the ground. The branch broke right where moisture had previously seeped out over several months. The break revealed a black rotting growth inside, under the bark, near the center of the tree. It was heartrot. Some hurts are more than skin deep.

When I returned home, I heard a commercial on TV. The announcer said, "How do you know when someone in your family is in trouble? Does a family member withdraw, have severe mood shifts, and sleep a lot? Call this toll-free number." I remembered the tree doctor saying, "Oaks do that sometimes." I thought how often problems are dismissed when they're little and easier to solve. Life is full of incidents in which a person hears, "No, dear, everything's fine," yet deep inside one knows it isn't. How often has a parent heard that a child's behavior is no big deal because it's merely a stage that kids grow out of?

I had worried about the tree, yet I didn't trust my instincts enough that something was seriously wrong. I assumed because I was not an expert, that others knew more than I did. The oak had "wept" for a long time, yet, like so many clues, the problem was glossed over, ignored, neglected because the tree kept most of the what was bothering it deep inside. I thought about people that suddenly remember childhood trauma in adulthood. My childhood friend, a therapist working in the psychiatric unit of a hospital, had once said, "The subconscious mind won't reveal past traumas until the conscious mind is strong enough to deal with the pain."

I didn't know the tree had heartrot. In life, it isn't the visible manifestations of wounds that are worrisome; it's the invisible ones, the old hurts running deep inside that no one talks about. Those are the ones that eat away at the organs.

The signs are always there, if we are willing to trust our instincts.

Chapter 25

THE RING OF CARE

My neighbor Tom endured much as a result of the breakup of his marriage. Like the cork oak in the previous chapter that lost a giant limb, he felt as if he'd lost a part of himself when he lost his wife. He didn't sleep, and didn't eat. As the months past, I watched the tree mend, and found in nature clues about how to approach situations of loss.

Where the oak's giant limb broke from heartrot, it left jagged, dangerously sharp pieces of raw wood—like raw emotions—with dozens of crevices for insects to get inside the tree, and weaken the tree further. A city worker arrived at the cork oak and with a chain saw smoothed the break, so there were no ugly, sharp, visible reminders of the split.

Tom's stress led him to get bronchitis, and despite antibiotics, his immune system was weakened, causing the congestion to hang on for months. It was suggested to Tom that he take all reminders of his wife and put them in a box, seal it, then put it in the attic to be opened when he could look at the items as memories of happy times. This he was reluctant to do, at first.

The tree didn't seem to heal immediately, but after a few months, it was clear the oak had been working at it. Recovery required slow, yet daily progress lasting over two years. First, the oak started growing the beginnings of what would become a new limb, a branch that could restore its balance. It was clear that nature doesn't ignore a problem, instead it deals with the issue at once. However, the new branch didn't "rebound" where the old limb had been. Instead it immediately started growing a new branch a quarter turn around the trunk. New growth in a new location seemed to affirm that the old limb was gone and it wasn't coming back. New growth was not going to thrive where the old limb had been. A new beginning was needed. The giant oak made getting back to a well-balanced state a top priority.

It seems trees have an advantage over us, as if they have encoded in their DNA a message that says, "No self-pity allowed, living is paramount." In contrast, we have powerful emotions, which can muddy our decision-making and delay moving forward.

On the side of the oak where the huge broken limb left a deep scar, much was happening. Each day, the bark around the scar began to lift a tiny bit. Raising the bark allowed room for a new material to glide slowly out from between the bark and the heartwood. The new material, resembling the color and shape of a plain doughnut, began to encircle the wound. In time, the puffy doughnut-like ring sealed the vulnerable area, encapsulating the wound in a protective ring of care.

If the tree had fallen down, instead of losing a limb, a different scenario would have taken place. When a tree falls, it is not only uprooted—severed from former connections—but also has fewer defenses against all kinds of ants, bacteria, mold, and termites, which get in under the bark and start eating away at the heart of the tree. Similarly, if we let ourselves stay down too long, sadness and anger can eat away at us, adversely affecting our health and spirit.

Trees obviously can't pick themselves up on their own as we can. However, now and then we find a real warrior tree that has fallen, yet is still attached to its roots. It doesn't

Oak along Route 128 near Cloverdale, CA

quite fall horizontally. Part of it lands at an angle leaning on another tree. The determined warrior tree fights to grow shoots that become a new tree and life begins again.

Nature, as teacher, shows us to call on our own inner resources, then along with family and friends build a strong circle of care after a painful event. When we face a difficult, troubling ordeal, talk about it and try to make peace with it quickly. Otherwise it festers and will get worse. Lastly, fallen trees remind us not to wait until we are down to find help, because then it's much harder to get back on one's feet.

In human terms, Tom also built a ring of care. He half-heartedly went with friends to Habitat for Humanity. There he soon engaged in a physically strenuous activity that let him literally hammer away at his anger and sadness. He also found a cause that let him put his hurt in perspective. Although his wife was gone, he still had a roof over his head and a career. He again felt good physically. Within eighteen months of his wife's leaving, he had made many new friends; one, a lovely fellow volunteer, got his heart beating fast again. Tom's work with Habitat for Humanity became part of his ring of care for himself, and in the process he helped build a house for a single mother with three children in need of a livable home.

In your neighborhood, you may notice the permanently visible "doughnuts" on trees where a broken limb was cut. I think of those "doughnuts" as the tree's badge of self-love for healing itself. We have the same assistance to give ourselves because we, the other Standing Ones, have been lovingly endowed with similar capabilities and strength.

Chapter 26

NEW YORK CITY

This is the story of a remarkable man who built a human ring of care that included strangers.

I was on a business trip to Manhattan in January. As much as I love that magic city and its incredible people, I felt out of sorts as I stood in my friend Alexander's apartment. With my nose pressed against the window, I looked down onto concrete streets, brownstones, glass, and steel. Cold rain dripped off the roofs and ran down the gutters to the sewer.

Alexander asked, "What's bothering you?"

I turned to him, a big bear of a man. "Coming over here, between the Met and your place, not once, but twice, in broad daylight, I saw spaced-out kids shooting up. They weren't much older than my own children." I struggled to shake the sadness I felt so that we could have a pleasant visit, but first we talked about the terrible toll drugs take on people and their families.

As we spoke, my eyes scanned the street below. It was then that I realized something and blurted it out. "There are no trees on your street."

"No, there aren't."

"I have such a longing for the peace I feel in the oak grove."

He smiled, and kidded, "Spoken like a true Californian."

I grinned. "The ache to walk among the oaks is a surprise to me, too. Those spaced-out kids' eyes keep flashing in my mind. I need to connect with nature."

With the drama of an opera singer that he was, he picked up a potted plant. "I don't have an oak for you," then, with a twinkle in his brown eyes, he bowed and handed me a potted jade tree to hold as consolation.

I grinned, thanked him and then wrapped my arms around the jade. "Ever notice," I asked, "that leaning against a tree helps you get grounded?"

"No," he replied, contemplating the possibility. "Why do you suppose that happens?"

"Maybe it's the roots going deep into the earth that helps us get centered."

He nodded. "I know this city needs its Central Park." A faraway look filled his dark brown eyes. "Sometimes I long for the smell of the magnolias in Charleston." He understood.

The doorbell rang. A voice from downstairs informed him a delivery had arrived. I heard Alexander unlock the multiple bolts on the door. Click, click, click. He went downstairs and signed for the food delivery.

While he was gone, it struck me that people install locks on their doors. In contrast, nature has many doorways—entrances to a woods, caves, canyons, mountain passes into valleys—but the entrances have no locks. Mother nature doesn't barricade herself away in fear or arm herself against man. Nature is as open as a meadow and honest as the seasons.

In New York, dwarfed by skyscrapers, I hadn't seen the sun setting on the horizon to know where I was in relation to the land. Surrounded by concrete and glass, it facilitated forgetting that all the people on the crowded streets are connected to the beauty in nature. Living as I do with open spaces nearby, it's easy to see nature's thread running through life. The thread weaves in and out with every wildflower and bird in flight. In the oak grove nature's beauty and generosity becomes tangible as all that lives within the canyon adds a thread to the rich tapestry of life.

I wondered if I would see the world differently if the only reflections I saw came from mirrored glass and shiny steel. Rain runs away from concrete just as hard hearts don't cry. Outdoors, mother earth quickly absorbs the rain that falls like tears.

The next day, nasty winter weather continued as I waited in SoHo for a taxi. When I got one, the driver drove like a madman, away from uptown where I asked him to take me. He explained that with the one way streets the route he was taking was faster.

"Did you come to New York alone?" he asked.

Something in the way he asked the question raised the hair on my neck. Before answering I looked at his eyes in the rearview mirror. They appeared wild, glazed and crazed on drugs, I suspected. His questions became more personal.

"Do you have any friends here in New York?"

"Yes, I do."

"Do they know you're in town?"

That was the question that got to me. When he turned onto a crowded one-way street, he cussed the congestion and became intensely agitated. It was a reaction out of proportion to the situation. Then the light ahead turned yellow. We were several cars back from the intersection. As he cussed to a stop, I reached in my purse for my wallet. I glanced up and saw him watching me in the rearview mirror. The meter read three dollars. In my wallet I only had two singles. I took out a five. He lifted his hips off the seat and twisted his body so he could look down into my open wallet. His eyes held steely on mine. He barked an order, "You'll pay me when we get where I'm taking you."

I threw the five on the backseat and lifted the door handle.

"What the hell?" he yelled reaching his arm over the seat towards me.

I escaped from the cab and quickly maneuvered through two lanes of stopped cars. At the sidewalk, I took off running in the opposite direction to the one-way street the cab was headed down.

It started sprinkling. The dark sky overhead looked like a cloud burst was only minutes away. Traffic was a mess. I crossed the street, ran another block and luckily I got another cab quickly. I told the new cab driver what happened in the last taxi. When I reached my destination, I got out, and still shaken, I grabbed my briefcase, but left my purse in the second cab. Money, credit cards, everything was in my purse.

The second cab driver learned I'd left my purse in his cab when he picked up his next passenger. He looked through my bag in an effort to locate me. By calling the emergency number in Los Angeles, he tracked me down in New York City. Later that night, he called to say that he had my purse. I was so happy and grateful that I asked him if he would pick me up the next morning and drive me to JFK airport.

When we arrived at the airport, he took my luggage out of the trunk then a porter took the bags. I paid the cab driver for the fare then told him I wanted to reward his honesty. I literally emptied my wallet—a little over eighty dollars—into his hands and to the cab driver's protest, saying it was unnecessary.

The porter stood by, watching the exchange, an expectant expression in his eyes for his gratuity. "Oh no," I muttered, "I have no money for the porter." I looked fleetingly at the taxi cab driver. He laughed and peeled off two fives.

"Here's one for the porter, and one for you so you have some money in your pocket."

The porter motioned another porter. "Come here. You've gotta see this. The cabby's giving the passenger money and it ain't change."

Fully aware of the porter's amusement, I told them of the taxi cab driver's kindness. The passengers standing nearby listened. Soon the crowd was applauding the honest taxi driver.

It turned out he came from Columbia, South America, and he'd been through enough pain and cruelty in his native country that he promised himself that, if he got to America, he wouldn't add to anyone's heartache in his adopted country.

In the driver of a bright yellow taxicab, I found beauty as heartwarming as a walk in the oak grove. Just as hillsides after a brushfire return to green in spring, my spirit felt centered again thanks to his integrity. While in New York, I never did get to hug an oak tree, but as I hugged the taxicab driver in a final farewell, I knew that the grandeur in nature grew in the man he'd chosen to become.

Chapter 27

WOODPECKER'S BAND

As I walked along the canyon road toward the oak grove, I heard an unfamiliar noise filling the air. The tapping sounds produced a soft, gentle melody, much like a drum roll. Da, da, da, daaaaa. Da, da, da, da. With my every step, the tapping got louder. Da, da, da, daaaaa. I became aware that I was walking differently. My steps followed the rhythm of the music in the air.

Trying to locate the source of the drumming, I looked up into the oak trees. Then I noticed that overnight the ancient oaks had budded in bright apple green. Up in an oak tree, out of the corner of my eye, I caught a glimpse of a woodpecker's bright red head. Then more red appeared in that and surrounding oak trees. At least two dozen woodpeckers were scattered in a couple huge oaks.

The drumming drew other people who were out for a walk. We all stopped to listen to the woodpecker's band. At first I thought the birds were after a tasty morsel that accompanied the oak's new leaves, or they were carving nests, but that was unlikely since the woodpeckers were too close to each other. Then a more knowledgeable person in the crowd said they were sending out love signals to attract a mate. After all, it was spring.

I proceeded to walk down the street, took a left and headed toward my favorite oak tree. Two woodpeckers flew into the tree and started pecking. Rat-a-tat-tat, da, da, da, daaaaa. Since the oak's branches started low to the ground, I put my ear to the branch and discovered that the tone produced by the woodpeckers echoed in the limb. I put my ear to another branch only to learn the echo varied with each branch. One branch had a deep, resonant pitch, yet when I pressed my ear to another limb the pitch was higher, the notes shorter. I guessed the tone depended on length, volume, and density of the limb.

Little yellow and gray finches landed in a nearby tree. Their sound was sweet. Between the woodpeckers and finches there was a duet.

For a few moments, I imagined traveling back in time until I saw myself living in a prehistoric age. In my reverie, others and I gathered near the entrance to our cave. We sat in a circle around a campfire after the sun went down. For amusement, we attempted to replicate the music of the birds we'd heard during the day. One of the men tapped on a nearby hollow log. Someone got the idea to seal the ends of the log with hide the way a limb is sealed where the limb joins the trunk. When the drum was finished, a stick was used to imitate a bird pecking on a limb, and then the drumming began. Voices rose to imitate the warble of the finches. A song was born.

I don't know which came first, music or language. All I know is the sounds of nature awakened the music in my soul. It was a marvelous, magical moment, available to all that stopped to listen to the woodpeckers' band.

Spring and all its glory had definitely arrived.

Chapter 28

FISH FOOD

One Easter Sunday around six in the morning, I went outside to the garden to hide over a hundred plastic eggs filled with assorted surprises. I wanted to wake everyone, I was so eager for the Easter egg hunt to begin. A second cup of coffee later, I looked in on the sleepyheads and decided to walk the dogs.

Over the weekend, the oaks had bloomed. Up in the trees, the woodpecker band was playing again. A whole year had passed since I first heard them play. I was so happy at the wonder of hearing the "band" again that I skipped a little down the canyon road, rounded the bend, and turned onto the greenbelt.

Regretfully, I noticed the cork oak tree was "crying" again. Only this time it was sobbing. On the far left-hand branch, the bark was deeply cracked. I paced off the length of the limb that could fall in a windstorm, and calculated about thirty feet. If the limb were to break, then the giant tree would have further shrunk in size. Then, as I scanned the tree with alarm, I spied another wet spot beginning in the very center of the trunk.

I stood there saddened. It was Easter, the day of the Resurrection. "Please *live*," I prayed. The symptoms were clear. The tree had more heartrot. Remembering the last windstorm, I thought that the diseased limb might break and hang like a loose tooth.

Even if over time the oak were to grow another equal size limb, the heartrot in the center of the tree caused me much concern. Then I remembered that my mom treated ailing plants with fish emulsion. It was worth a try. A little emulsion, full strength, no diluting the smelly stuff. I was convinced it would work since a considerable amount of the greenery on the planet grows in the ocean near the shoreline in kelp beds, and fish are the ocean's fertilizer. Then I encountered a snag. At the local nursery, the owner no longer stocked fish emulsion, but he would special order some. It would take days to arrive.

Impatient to help the ailing oak, I remembered growing up that the house with the prettiest roses belonged to a woman those who planted fish heads and tails in her rose garden. On Monday after Easter, I talked to the man down at the fish market, and he agreed to save fish scraps for me.

The next day, I picked up the scraps and headed for the oak tree armed with fish heads and tails in a brown paper sack. I planned to bury the fish in the ground rather than a plastic bag, which I'd feel obliged to carry home and put in the trash. I was ready to accomplish my mission. I trudged down the road at noon on a hot afternoon with a mini-shovel. I felt the fish leaking through the bag and smelled the aroma on the wind.

Two cats start following me. Meow. Meow. I should have brought our dogs.

When I got to the tree, I shooed the cats away. The fish heads were buried, for the moment, depending on the crafty cats. I hoped the fish heads would work their magic. If not, when the fish emulsion arrived, I'd pour it on the tree in the spots where it appeared sick.

I had to do something. The tree was fighting for its life!

When we help a tree live, a river heal, a wounded animal recover, or provide food for dwindling numbers of songbirds, in the process we affirm the sacred circle of life and our role as its custodians.

Chapter 29

LEAVING SOMETHING BEHIND

Sometimes one incident serves to magnify another. Such was the case when the oak tree was dying. At the same time, my mom was in the last stages of cancer. What upset me was that in both instances I felt powerless to change the outcome. Just as I often sat by the oak tree to think, I often sat by mom's hospital bed in the evening while she slept. In quiet moments at her bedside, a host of childhood memories returned, including the first time death touched me.

I was about three years old when my best friend's father died. I remember the shock of answering the door to a policeman, who asked to talk with my mother. He came right into our house. Mom invited the policeman to sit down at the kitchen table. Over a cup of coffee, he explained that the lady next door, whom I called Aunt Gen, was going to need a friend because her husband had died instantly in an automobile accident.

Aunt Gen's son Jack was my age. He was my best friend. Mom and Aunt Gen decided I should stay overnight with Jack. At such events, it's the little things that stay in the memory, like his mother's rose chenille robe, and the moon shining in on us that night. I was there to comfort him, and he rewarded me by letting me take the top bunk of his bed.

The next month my grandmother died.

About this same time, I took to dressing up as a gypsy in a bright red sweater with layers of skirts. I added many strings of colorful beads, but most important was a jeweled broach that transformed me into a gypsy or so I thought. Dressed in my multi-hued clothing, I climbed onto the living room desk by the front windows so I could look over the treetops. Atop the desk, I'd sing and dance much to my mother's objections. She feared I'd fall and break a bone, but I insisted I wouldn't. I told her it was extremely important that I stand up high because I was waiting for a tall prince. He had gone off to fight his mean uncle. Being so short, if I didn't dance on the desk, he wouldn't see me when he returned from someplace far away. At this faraway place, I imagined people spoke a language that sounded like the Polish family down the street.

She shook her head with dismay, trying to understand where I ever got certain notions. Mom wasn't fond of gypsies from an incident in her youth. After she talked to dad, they forbade me to dance on the desk, taking away both my dream and new game. I felt deeply hurt. If I couldn't dance like a gypsy on the desk, I wouldn't see him riding toward me on his horse. He might never find me. In my heart I had promised him I would make it possible for him to find me.

I remained upset for weeks. One day, while cutting out paper dolls, I crawled behind a chair and cut a section out of mom's new lace curtains. I hid the piece under the rug. Months later when the rugs were rolled back for spring cleaning, I got my first remembered spanking, not only for cutting the curtain, but also for denying doing the deed. That someone else must have done it was a bit of a stretch considering I was an only child at the time.

In the hospital room, when mom awakened I told her I was sorry about cutting her curtains. She'd forgotten all about the incident, which was her way to let "bygones be bygones."

A few days later, after another visit to the hospital, I drove through Old Town, the site of many quaint shops. I stopped to browse. Inside one store, the inviting scent of eucalyptus and orange peel filled the interior. Across the room, stood a marvelous antique Beidermeir armoire. The rich patina on the ash and oak double doors was amazingly beautiful. The armoire dated back to before the War of 1812, when Jefferson and Adams were still alive. Perhaps, it might have belonged to the tall, blond prince from my childhood fantasy. My fingertips touched the satiny wood as I traced the swirling movement of the graining.

I momentarily wished that I could afford it, to keep as a family heirloom to pass onto my children. Then I remembered my mom had asked what I wanted of hers as a keepsake. I was at a loss for words. The things she'd leave behind weren't as important as she was. How could her favorite vase, lovely though it was, replace her? I would give a thousand such vases if she could continue living, for she was a good-hearted woman who loved well.

As I gazed at the armoire's mellow ash and oak, I thought of all the beautiful woods used to make the antique furniture in the store. Many pieces were made by cabinetmakers that used woods from the pristine untouched forests of North and South America. The old growth wood represented a tree's way of leaving something remarkable and enduring. The best gift my mom would hand down was expressed in the words of John Lame Deer of the Lakota: "Love is the only thing you can leave behind that's permanent. It's that powerful."

Since love is powerful, perhaps once upon a time in a far off land a gypsy and a prince really did care for each other.

Chapter 30

GIVING BACK

After my mom was buried, I would walk through the oak grove thinking about her funeral. My brother did an outstanding job attending to all the arrangements. He selected a feminine, pearled blue-gray coffin with pink satin lining. It was mom. It was lovely, and she looked beautiful and healthy although she weighed only sixty-eight pounds when she died.

Before she died she told me she liked a particular dress and had felt good wearing it. It was a sheer multi-pastel stripped dress with long sleeves that topped her list of summer favorites. I mailed that dress along with the matching satin under-slip to the funeral director in the Midwest.

Later that day my dad came over to my house and pestered me with questions. "You did mail your mother's clothes to the funeral parlor, didn't you?"

"Yes, Dad."

"You're sure?"

He looked exhausted. "Very," I replied patting his shoulder.

"I dreamt about your mother last night. She kept me awake half the night, bugging me about what she's wearing." He scratched his head of gray hair. "Any idea what's upsetting her?"

"Not a clue," I replied. Since dad was clearly distressed, I took his dream seriously. "I'll think about it," I promised.

While walking in the canyon the answer dawned on me. It was a mother thing, by a mother who had never burned a bra in her life. Mother was very modest. I'd forgotten mom's bra and panties. With understanding and a smile on my face, I rectified the matter in a hurry. I went to Victoria's Secret, where Mom had never shopped. I got her something *special.*

When I told dad the story, he laughed. The next day he reported a full night's sleep.

After the church service, dad and I were talking to the funeral director. My father found it heartwarming when the funeral director said that mom would look as pretty fifty years from now as she did that day.

"Fifty years!" I remarked.

The director proudly pulled his shoulders back. "Yes, we now make caskets that are hermetically sealed so air does not get in to deteriorate the body for years to come."

Dad nudged me, smiling brightly. "Isn't that something?"

"It certainly is," I replied.

The director went on to explain that they would put the coffin in a concrete vault and seal the concrete lid. The vault would get dropped into the grave in the ground then covered with grass. I later learned the vault would help prevent the ground from compacting and sinking, which would make mowing the lawn more efficient. I wondered if a hermetically sealed metal coffin buried inside a cement vault was a sign of progress. Or, was it symbolic of how far we, as a people, have removed ourselves from feeling one with the earth?

Granted, how a person rests in peace is truly a personal matter. I thought of my own going. Funerals make you think about things like that. While standing at the gravesite, I knew I wanted something else. I fell in love with this planet the first time I saw how it looked from space. I'd like to be buried directly in the ground in a wooden box. After I'm dead, that's the last and only possible gift that I can give back to this dear earth, in return for all that it has given to support my life.

I'd like to participate in the natural cycle of life by becoming fertilizer for a spreading oak tree with a bench beside it. The kind of shade tree where birds come to sing, and people sit on the bench on a hot day, the way I sit beside the oak.

Chapter 31

A MATTER OF OPINION

In the canyon my favorite wild oak started from a little acorn hundreds of years ago. Two thousand miles away in the Midwest, a wild American elm also grew from seed. A major difference in the two wild species is that the oak has remained secluded in a canyon while the elm became part of a quiet Midwest neighborhood.

It was estimated that the graceful elm started growing in the 1860's, about the time of the Civil War. When the elm began the area's land use revolved around farming and woodlands, but eventually homes were built creating a neighborhood. Thereafter, the elm shared space with a Cape Cod bungalow. This is the story of what happened to the elm and the people who live in the bungalow.

One warm spring day when bright, green buds held the promise of spring, my childhood friend called me on the phone. I could tell by her shaking voice that she was upset. She said, "I'm looking outside for the last time at that beautiful old elm in the front yard." Despite her protests, the graceful elm would meet the city's chain saw the next day.

"That's sad. It's such a graceful elm. Is it diseased?" I asked.

"Oh, no," my friend angrily replied.

With disbelief, I said, "It's not infected? But, there's so few great elms left. Why is it being cut down?"

"My grandmother insists on destroying that tree because *'it's messy.'* Between working full time and going back to school for my masters, I don't have time to always pick up fallen branches or rake the leaves."

"Well, elms are not pack rats," I replied trying to see both sides. "They do have a habit of cleaning out unnecessary branches. Is the mess that serious a problem?"

"Only my grandmother thinks so," she lamented. "I like that tree. It makes me so livid to destroy it." In her tone anger was unmistakable. "By summer I'll swelter in my upstairs bedroom without the elm's shade."

For over a century, the giant American elm stood in the grassy area in front of the house along with other equally old elms that graced the neighborhood. All the elms started as wild trees long before the houses were built in the 1940s. When the developer built the houses, the city planners respected the grove of elm trees enough to take out some and leave others, then plan the streets to accommodate the trees.

The elm's elegant branches extended like a big green banner from one side of the property to the other. Its majestic height dwarfed the home and lot. The elm's limbs acted like ceiling fans, moving the Midwest's hot humid air, and shading the house to keep the non-air-conditioned dwelling about ten degrees cooler during the hot, dog days of summer.

Its roots didn't attack the home's foundation, or rip up the driveway because when trees begin from seed out in the wild, they develop longer taproots than planted trees. For instance, a little foot-high oak sapling may have a taproot that is much longer than the tree is high. Since wild trees can't rely on sprinklers, they send out a taproot to tap into underground streams. On the other hand, trees planted in a yard with sprinklers are a bit lazy. The grass needs lots of water so the trees don't need to shove their way past and through rocks to get to an underground water source. In contrast, wild trees go on a search and seize mission to find enough water to sustain their enthusiasm for growing.

My friend asked me to talk to her grandmother. When her grandmother got on the phone, she was unwavering. "That tree's a damn nuisance," she stubbornly complained, as if cursing the most vile creature on earth, "and, I want it gone." Then she gave me what she believed was the deal clincher, the one irrefutable argument that no sane person would argue against. "I told the city the old tree was a danger to my property, to neighbors, and I'm too old to take care of it. Therefore, the city by law has to cut down the elm for free." She came up with this plan after talking to her cronies at the Senior Citizen's card club. "It'll save me hundreds of dollars, otherwise I'll have to have that tree trimmed," her grandmother stated. Having lived through the Depression, she'd always been a practical, very thrifty lady. She thought it immoral not to take advantage of any money saving opportunity.

"You see," my friend lamented, "no amount of persuasion will stop her from destroying the elm. I've argued that the elm's survival might hold clues that would help scientists develop a disease resistant elm, but I only wasted my breath." With a sigh, my girlfriend continued, "My grandmother thinks there are plenty of other healthy elms on the street and scientists could use one of those. It does not matter to her why I want the elm to stay, or that my upstairs bedroom gets beastly hot in the summer."

When I asked what her grandmother said to that, my friend fumed, "She told me I could buy myself a fan."

Her grandmother got on the phone again. Determined as ever, she said, "For whatever amount of years I have left, I don't intend to live them raking leaves or picking up fallen branches." Then she added, "And, I don't intend to wait until my granddaughter gets around to raking the leaves!"

The following day the city's trucks arrived. Men with chain saws, ropes, and hydraulic ladders arrived to surgically remove the limbs before cutting into the heart of the tree. Years of growth turned to sawdust in a few hours as the chain saws gnashed and slashed. The city told her grandmother she'd need to order a truckload of topsoil to fill the hole in the ground where the roots once thrived.

My friend was very distressed when she called me the following evening. She had come home from work to find the lawn a mess and a hole in the ground as deep and big as a swimming pool.

"How does your grandmother feel now that the tree is gone?" I asked.

"She's inside the house now calling her cronies. Her delight has reached new heights because she got rid of a nuisance without it costing her a cent."

My friend had new reasons to consider it reprehensible to cut down the ancient elm. "And, get this," she seethed, "*I* have to pay for the truckload of dirt to fill the hole."

Both women had different opinions on that elm tree. One woman had several grandchildren when the environmental movement began. The other had recently returned to college to get her masters and was surrounded by new thinking on the environment. Just as the elm tapped into the underground stream, the two women tapped into their streams of thought—only they were different streams.

Chapter 32

IGNORANCE

I worked for a time in a building where there were no windows available to me during most of the day. Walls engulfed me. I missed the feel of fresh, quick breezes blowing across my face, the rustle of the wind through the trees and the songs of birds. After weeks of being closeted indoors, I took myself over to the greenbelt where the flock of crows had drawn me to the oak. I was sure that walking among the ancient oaks would renew my spirit. When I reached the greenbelt, things were not as I expected.

In the ivy, scattered near the cork oak's base was an inordinate amount of dog poop. I tiptoed around it. When I reached the tree, I rested my elbows on the oak's lowest branches. Right there, a few inches from my hand, on the crook of the oak's trunk, was more dog poop. I couldn't believe it. I felt violated. Who would put feces on a tree? I went to find a stick and leaves to clean it off.

A few weeks passed before I returned to the cork oak. I was standing there deep in thought when from the other side of the wall I heard a sliding glass door open, and a man's voice greeted a couple dogs. His footsteps echoed on what I assumed was a patio, then I heard the unmistakable sound of a pooper-scooper scrapping against concrete. Suddenly, a shovel full of dog poop came flying over the six-foot wall, coming right at me.

"Hey! What the heck are you doing?" I shouted as I ducked.

"Oh sorry," replied the unseen man. I heard his footsteps on loose gravel as he went to the corner of his lot adjacent to the open area where the ancient coastal live oaks grow. I could hear him continue to use the pooper-scooper. His footsteps crunched on the gravel as he started to walk away from the wall. I didn't anticipate that over the wall, at a right angle to where I stood another load of doggie-do would fly through the air.

"Hey, cut it out!" I hollered. "People could be standing here communing with nature, and you're throwing shit at them."

He didn't say a word. I heard his heels as he walked across his patio. The sliding glass door opened, then closed. It all happened years ago and never again did I step on dog droppings or find them on or around the oak. I applaud whoever lived in the house behind the wall for changing his ways. The face I never saw re-affirmed my belief that most people act largely out of ignorance, unaware as to how their actions effect the environment and other people, rather than deliberate malicious intent.

CHAPTER 33

THE ANCIENT OAK

From my former home I could see a beautiful one-story house that stood alone on a nearby hilltop. A private road led up to the house. From the hilltop home a magnificent view of the valley extended for miles.

In the front courtyard an extraordinary California live oak nearly a hundred feet tall and just as wide towered over the house, framing it in picturesque splendor. The majestic oak had looked out at the changes in the valley below for over 600 years. When Columbus sailed, the tree was already a hundred years old. During the Revolutionary War the tree neared its 400th year.

Throughout the centuries the stately oak shed its leaves, enriched the soil, and dropped acorns for squirrels. Each spring the oak blossomed about the same time the swallows flew back to Capistrano. The tree withstood the challenges of the centuries by enduring the area's numerous earthquakes, droughts, and brush fires. For many years the oak was one of many oaks on a ranchero. Then in the 1880's the oak began sharing its space on the hilltop with a house that was owned by one of the valley's founding families. Decades later another family bought the property and built the sprawling one-of-a-kind house next to the ancient oak. It must have been hard to build around the oak, but worth it in the days before air-conditioning. The huge oak shaded the home, cooling it on days when the temperature soared above a hundred.

A couple homeowners and a hundred years later, the house came on the market. Having admired the house for years, I went to the open house. Inside, a long living room extended about forty-feet from the front door straight ahead to a wide wall of glass. The walls on the sides of the room were covered with no ordinary wood. To call it paneling would not connote its beauty since there are so many different kinds of paneling. Imagine a series of two-foot wide by nine-foot-high butterfly panels of gorgeous wood. The wood, which looked like walnut, extended from floor to ceiling on both sides of the enormous room. The wood had darkened with age, but it still remained vibrant with mellow swirls of red and amber. The well-maintained wood caused me to rub my hand across the grain as if it was a flawless, muscular, strong back. It was beautiful.

A new family moved into the hilltop home. The new owner preferred light colored walls, so he had the decorator cover the wood with textured ivory wallpaper. The owner also disliked the ancient oak tree for numerous reasons. It deposited bushels of leaves on the driveway, which his gardener spent too much time raking. At night he heard acorns dropping on his roof like an annoying drip of a faucet. Early in the morning squirrels scampered overhead, crossing the roof to eat acorns for breakfast. He wasn't going to put up with the inconvenience of that darn tree.

The homeowner sought a permit from the city to bring in heavy equipment to cut down the oak tree. The permit department said no. It had rules to protect oak trees and that oak had to stay right there. The homeowner went before the city council. When he spoke he commanded attention. "You are limiting my right to cut down a tree on my own property." It was clear that, if he could, he'd cut down that oak in a heartbeat. "But, laws are laws," he conceded, "and

I'll respect yours." Yet, knowing his way around city hall, he got a permit to "trim" the old oak.

It was a curious thing. The man who originally conceived of developing the area into a residential community bought the land years ago. At one time, one of his companies held title to the house on the hill. He was so rich he bought land greater in size than the state of Connecticut in the Amazon rain forest for less than a dollar an acre. *Bzzzzzzzzz, Bzzzzzzzzz* went the chain saws to level the Amazon land. In cutting the rainforest, he devastated millions of acres of one of the largest gene archives on earth. What was destroyed was a fragile ecosystem representing 500,000,000 years of evolution. I saw it on NOVA at least twenty years ago. I suspected the beautiful wood in the house on the hill might have come from one of his forests.

The old mogul still owned land in the area. He wanted to build a multi-million dollar fifteen-year development project on over 129 acres of meadowland. In a small town fifty miles from Los Angeles, over a million square feet of new retail and commercial space spelled sprawl to the local residents. A lot of folks resented the density of the proposed development, not to mention the air pollution and subsequent freeway congestion.

Eager for ample new revenue, the city council indicated support for the 129-acre project. Local residents put up intense vocal opposition. Since re-elections for city council members were only weeks away there was hope that environmentally minded candidates would replace the outgoing members. But, late one night the existing city council quickly approved the project.

Opponents of the project were livid. They hoped they could over-turn the council's decision by putting a referendum on the ballot to repeal approval of the huge development. The signatures needed to put a referendum on the ballot were quickly gathered. On the last day the petition along with the required signatures were handed to the city clerk. The city planner rejected the petition on the premise that since the 700-odd pages explaining the ins-and-outs of the referendum were not directly attached to the petition when received, the citizenry may not have known what they were signing. Technicalities. Letter of the law.

The furious citizens asked for a ruling from the state's attorney general. Over the weekend, the attorney general ruled, siding with the city and against the citizens. When I signed the petition, the 700-page document was attached. I didn't read it, but I was clear about the petition's intention since it had been the talk of the town and on the newspaper's front page for days.

Regrettably, the time left for getting a referendum on the ballot had passed. The development would go through. Years later, when Southern California went into a recession, the new homeowner declared bankruptcy and sold the house. The developer of the 129 acres passed away. The commercial development that was built is thriving.

The history of the home on the hill connected it to the land developer, who thoughtlessly cut down a huge part of the Amazon rain forest. In contrast, the children that moved into the hilltop home are part of a generation educated to protect the environment in ways unknown fifty years ago. Of all the changes through the years, the one constant is the ancient oak. Like a silent sentinel in the courtyard of the house, the oak still stands on the hill as it has done for over six hundred years.

CHAPTER 34

THE SEED REPRODUCES ITSELF

Acorns are falling from the oak trees, littering the ground with possibilities. Yet, in a sense they are limited possibilities because acorns turn into oak trees. If you plant an acorn, you won't get a Christmas tree.

The earth is designed so that seeds reproduce themselves when planted in fertile soil. Likewise, the "seed" of intent reproduces itself because the human mind—like the soil of the earth—is fertile. The seed starts growing the moment the deed begins. At that moment action gives life to what was planted in the mind and heart.

Like the oak that churns out thousands of acorns during its lifetime, humans also duplicate their actions. A kind person is kind to animals, children, employees, spouse and strangers. A person with integrity does the decent thing even when no one is looking.

But, let's say the person isn't so nice. If someone does dastardly deeds—intimidation, mistreatment, physical harm, or any other abuse—it's unlikely the person is doing the same to you alone. Let's say the person is arrogant and continually talks down to you. It's quite likely the person has been equally condescending to the deliveryman, subordinates at work, as well as family members. The seed reproduces itself. There's a human tendency to blame ourselves, thinking the doer would stop, if we tried harder, or we were somehow different, better, prettier or smarter. It's not true. Abusive behavior is done to make the perpetrator feel more powerful by gaining control over others. When we encounter a person, group or company that plants seeds that are ultimately harmful, it's best not to give harmful seeds a chance to grow in "one's garden."

Panorama of oak covered foothills, Thousand Oaks, CA

This is not to say that abusive people don't change. They do. Like everything else in nature, we are able to adapt and mutate to create anew. However, the "seed" that we can control, which we maintain the right to modify, is ourselves. The freedom to change resides within each of us as a personal gift. There are within us many wonderful "seeds" waiting for us to claim them and let them grow.

Here are three true stories that demonstrate how seeds of kindness reproduced themselves.

In Macedonia in August of 1910, as summer neared its end, a baby girl was born to a contractor and his wife. From the time the child was little, she learned to care for those in need from her family, who regularly made it a habit to go into town and personally work with those less fortunate. When she grew up, she joined a religious order in Ireland and was sent to India. There she dreamt of a world where everyone had a home in which they were loved and cared for as she had had as a child.

On the streets of Calcutta, she saw the abandoned and destitute of different castes and religions lying on the dirty city streets waiting to die. She managed to get an old home to which she brought people off the streets and gave them a bed. But, she gave them something more, a place where they could die in peace, dignity, and surrounded with love. She wanted to show the dying outcasts that, whatever their religion, they were loved by God. In time, word of what she was doing spread. Other women eventually joined her.

When she was born no one knew that the infant girl would grow up to become Mother Theresa, and continue to plant seeds of compassion that she learned in childhood. In her lifetime she received the admiration of people around the world, the Nobel Peace Prize, and she became one of the greatest humanitarians in history.

In 1985, when singers John Mellencamp, Willie Nelson, and Neil Young were concerned about failing family farms, they organized Farm Aid concerts to assist financially troubled Midwest farmers. Years later when Willie Nelson got in trouble with the IRS for back taxes, the seeds of his own generosity bore fruit. When the IRS auctioned his possessions, fans from all walks of life raised money to buy his personal belongings at the IRS auction so he would eventually be able to buy back his personal treasures. The seed the musicians planted reproduced itself and led to a wide variety of aid concerts all over the world.

Fame and star power is not a perquisite to doing great deeds. This was the case on 9/11 when people in New York and the nation's capitol built a ring of care around the people in their cities. In the face of evil and hatred, ordinary people with extraordinary character sought to plant accord, compassion and unity. In New York City, Mayor Rudolph Giuliani and members of the fire and police departments became unforgettable heroes. Their actions are forever stamped on our collective memory for their courage and dauntless efforts in the face of tragedy. New Yorkers—strangers in many cases—befriended each other. Restaurants generously donated food. Hospital personnel worked tirelessly. City workers helped families find information about loved ones. From across the world, people sent money for victim's families. Countless others encountered evil and refused to plant more of the same, but instead chose to plant empathy and goodwill.

A couple years later the seed planted by New York City workers on 9/11 had grown and was harvested during the Blackout of 2003. Over 50 million people from Connecticut to the Great Lakes had their power cut on a blistering hot summer day. Millions trusted that all that could be done was being done. When New Yorkers responded with calm, resilience, and good spirits, they again affirmed the best in the human spirit.

Earth, just as with Eden, is a garden in which we reap what we sow. If we plant acorns, we get oak trees. In a similar way, if we plant meanness, we get meanness. Yet, if we plant kindness, we get kindness. If we give respect, we get respect. The seed reproduces itself. This is one of nature's greatest lessons.

Chapter 35

SQUIRREL

When Squirrel scampered across the lawn, I had no idea the world he was about to show me.

Squirrel quickly climbed a pine tree, plucked a big pinecone from a lower branch. He stopped and sniffed the air while his red tail twitched nervously. He spotted me then barked in a threatening tone.

I'd heard that animals become less skittish when they hear the scale, so I stood perfectly still, hands at my sides, and sang softly, "Doe, re, mi, fa, so, la, ti, doe." Squirrel stopped and stared at me. A few notes were off key so I sang the scale again. The second attempt seemed to work. Squirrel dropped down onto a lower branch until we were almost at eye level, about five feet apart.

It appeared that he figured I was okay since he chattered calmly something indistinguishable as he made himself comfortable on the branch. Holding the pinecone in his hands, he carefully examined it by rotating it with his little hands. Then, reminding me of a carnival barker, he lifted the pinecone in the air. Squirrel looked at me as if to say, "Step right up, see the best trick with a pinecone this side of the Mississippi."

Then Squirrel held the pinecone as if holding a rolling pin. He energetically rolled the cone back and forth on the tree limb. Then he struck one side of the cone hard against the branch. Tiny seeds encased in gossamer pouches floated onto the ground. Squirrel sat back and munched on the pinecone. Upon returning to work, he rubbed the cone some more, shook it, and again extended his arm, thereby encouraging more seeds to fall. And so it went: pine nuts for the birds, some for Squirrel, and the rest for mother earth to grow new pine trees. When Squirrel finished, the pinecone looked like an apple core. He dropped the eco-friendly core beside the tree.

Instead of picking a nearby pinecone from the lower branches, Squirrel scampered to get a pinecone from high up in the tree. Later I learned that on some pine trees the cones containing female seeds grow on the lower branches, the male cones are on top. So it was that Squirrel played a role in pine tree reproduction by mixing male and female seeds together on the ground.

Squirrel continued planting for the future, aware on some level that change was inevitable. Whatever shall come, Squirrel had dealt with the basics of life of which reproduction, food, and shelter were high on the list. There was nothing helpless about him. His motivations wore work clothes. When Squirrel finished, pine nuts encased in thin, gossamer pouches lay strewn on the ground. The sun will slightly dry the seeds then the wind off the ocean will lift the seeds and carry them to new areas.

Squirrel seemed proud of himself as he arranged his red tail to make it fuller, grander. Before me was a little squirrel filled with life, causing life to renew itself. "You're beautiful," I told him.

It may appear that Squirrel has it pretty easy compared to most humans, considering he has free food and shelter, plus no concerns about mortgages, taxes, no worry about the stock market, or reasonable treatment from his HMO. However, Squirrel works hard. For every thing that Squirrel takes from the land, he gives back to other creatures a goodly portion so nobody goes hungry. Squirrel may plant thousands of seeds in his lifetime, and if only one pine tree or one oak germinates then Squirrel's efforts are returned a hundred-fold every year.

After Squirrel finished eating, he sat back against the trunk, rubbed his hands, rested a few minutes then ran down the tree and into a hole. I figured Squirrel was tending to his other job, munching on the fungus on oak tree roots that he likes to eat, and by removing the fungus the oaks can get more nitrogen from the soil. The nitrogen helps the oaks stay healthy, which is vital. California's oak trees are critical to the state's ecosystem since they support over a hundred species of birds, mammals, insects, and deer.

Every time Squirrel took something from the earth, he took only what he needed, and made an investment in the future, not only for his own kind, but also for all the other birds and mammals that live near him. One squirrel doesn't plant an oak forest any more than one person planting a fruit tree eliminates world hunger. Yet Squirrel shows us how simple caring for the earth and each other can be if each of us takes care of the area where we live. A "squirrelly" world might be a pretty nice place to live.

CHAPTER 36

AN EARTHY WOMAN

Ever since tall ships sailed across the Atlantic bringing supplies to the burgeoning North American colonies, oak trees have stood like silent observers to the massive changes that continue to take place across this land.

When the colonists arrived, they saw a land of infinite possibilities. In letters home to relatives in Europe, colonists told of a healing land where the water and air were so pure they possessed curative powers. In thick forests lived a profusion of wild game along with rivers brimming with fish that gave testament to the natural abundance. In a world without telephones or television, word of a miraculous land spread like wildfire. In the bustling port cities from Boston to Savannah, people spoke of the land as if she was a lovely, generous woman.

Although the land beckoned with tempting possibilities of adventure and riches for many, America remained like a princess in the tower. She was not easily accessible, waiting as she did on the other side of a perilous ocean. Yet, in the mind's eye, she exemplified a young woman with lush untouched forests and mountain peaks to scale and climb. It didn't matter that the she was wild and untamed because she seemed to echo longings to be explored. She tantalized men who wanted to claim her, subdue her, and show her their "civilizing" hand.

Men of very modest means poured into the colonies, especially to the Tidewater area around Virginia and a smaller number to New England. They came by the thousands, lured by the promise of a small plot of land. The price they paid for free passage was high. They signed a contract for up to seven years of labor for an unknown master. Once here indentured servants received minimum food, shelter and clothing, but they often didn't receive the bulk of their meager pay until the end of their service to prevent them from running away.

In the Tidewater region, they frequently worked and slept with slaves; if they had relations with someone of another race, the law permitted extending the length of their indenture. During their term of servitude, they could be whipped or sold. When their years of servitude ended, they received their back pay and were eligible to buy a small plot of land to grow their own food.

Servitude was tolerable because the land offered something commoners found irresistible: the chance to realize previously impossible dreams of being their own boss on their own property rather than as a tenant farmer. America excited the human spirit with hope, thrilling it with the possibility of fulfilling every fantasy.

Thousands of men stampeded to make America their own, but far fewer women came as indentured servants, fearing being forced into prostitution. Because men far outnumbered women in the colonies, many men upon being freed married Native Americans from neighboring villages. However, the marriages were rarely recorded because the English banned such unions. Nevertheless, people are human, and men and women do find each other.

Marriages between English and French colonists to Native Americans sowed the seeds of a more compassionate way of looking at the land and so much more.

Two different philosophies met around the kitchen table and in the bedroom. One valued liberty and wore fringes and long hair to feel the free wind—the Breath of the Great Spirit—while the other spouse was still subject to the monarch's will. The independent, intensely freedom loving Native Americans—whose chiefs still are humble servants of their People—were a thorn in monarchy's side. A thorn to be removed because Native American ideals posed a threat to everything monarchy stood for back then.

Despite bans against mixed marriage, the proliferation of marriages that blended two cultures continued on the frontier and in areas under French jurisdiction where mixed marriages were not forbidden. French territory eventually became all the states that surround the Great Lakes and down the Mississippi River to Louisiana. Those marriages helped sow the seeds of respect for the land and new thought. Native Americans believed so passionately in equality that their unique, democratic views helped break down the rigid social pecking order common to monarchies at that time. For instance, Native Americans didn't sit on thrones, but sat in a circle where no one was above another. That example of equality opened a window onto the revolutionary idea that *all men are created equal.*

Yet, the Old World perspective that the landed gentry and merchants brought with them, including prejudice against other races and the right to have power over other people and the land, had very deep roots not easily extracted, even after the Revolution. Just prior to the Revolution, all but Pennsylvania and Maryland had become royal provinces run by men appointed by England's King George III. The appointees had put in place laws that cast the land, commoners, women, Native Americans and slaves in legally exploitative positions. Sweeping changes to those laws came slowly. Indentured servitude continued for another seventy-five years beyond 1776, and ended about the same time as slavery did.

In 1776, most states' laws held that a wife and the land belonged to her husband. Since women could not vote, they played no direct role in electing representatives who made laws affecting their economic stability or the philosophies behind the laws. They had no say in the enforcement of law. In some states, a wife could not even own property in her name. A woman's surest way of securing her economic future was to marry well. If a wife worked, her wages often belonged to her husband, just as the resources from vast tracts of land belonged to the titleholder.

Since colonial times, when the land was pristine, healthful and abundant, the land has suffered much abuse. Slowly, deliberately, she was robbed of her resources, dammed, mined, stripped and raped. Many women met a similar fate. The beautiful land served not as something to love, as a woman is loved, but as an asset for personal gain.

The settler's approach shocked—then broke the hearts—of Native Americans and their descendants. Each tribe or nation had its own land, but they erected no fences, so that people and animals could move and be free. They never conceived of *owning* or possessing mother earth. A mother was never for sale. Since only a woman can give birth, what came from the earth was inherently born of woman. Her "milk" was a gift from the Great Spirit to feed all the creatures that lived on the land. Earth's children's duty was to care for their mother. Such a view entails a spiritual connection and gives purpose to life. For Native Americans, an equally important Father Sky, representing the sun and wind, balanced the feminine nature of the earth.

Such thinking was promptly called pagan and aligned with witchcraft; stamping out such thinking became the self-imposed goal of the Puritans. The Puritans made up a large percentage of the population in colonial times and in the new Republic so that their opinions held sway. For other Protestant sects, a universal mother most likely appeared too close to the Catholic's universal Mother Mary, which was a concept the Protestant Reformation eliminated during the hundred years before the Puritans landed.

The belief that one can do with the land and the animals as one pleases prevailed. Old attitudes linger as long as they find acceptance and support. It took three hundred years—from the time the Pilgrims landed in 1620, until the 1920's—before women got the right to vote to choose their representatives in all local and national elections. That same decade Native Americans became citizens! Then fifty years later, in the 1970's—350 years after the Pilgrims landed—the environmental movement started in the United States and gained momentum around the world.

We now know that what one does to the land that one *owns* affects the health of neighbors and one's community. We've seen the rapid extinction of species, global warming that affects weather worldwide, and the effect of pollution on our own health. Mother earth is like every woman. She has her limits. At some point, she must and will fight back. This lovely land, this "desirable woman" that the colonists praised hundreds of years ago, has given generously to all those who came to her shores, and we wish mother earth to continue to do so.

There is much reason for optimism. A growing respect for the talents of wives and daughters and respect for the land continues due largely to millions of unsung individuals. The unsung are the ones that make changes in their homes and businesses through recycling, the products they buy, and values they endorse. Regard for protecting the land runs deep in the national consciousness and forms a growing ground swell to the extent that Americans list the environment among their chief concerns. Each decade has brought us greater understanding.

Something even more significant and hopeful is happening. The present generation is making impressive strides in creating deeper, more honest, and emotionally open relationships between men and women. The seed of mutual appreciation in relationships is an extension of caring for the earth because the more respect and understanding we, men and women, have for each other as human beings, the more respect we'll demonstrate for all that grows, swims, flies and walks upon the earth.

A bright new day is coming.

Chapter 37

WITHOUT ONE, NO OTHER

One of poet John Dunne's most famous lines is: "No man is an island, entire of itself; every man is a piece of the continent."

Occasionally, while walking in the oak grove, I'll notice clouds moving in the jet stream and think of Dunne's words. Then I'll reflect on larger weather patterns, becoming aware that the sun shining on the oaks is playing on the sea ten miles away. The same sun is reflecting off the needles of pine trees further north near Yosemite. Under my feet, oak roots hidden below the ground reach downward into underground streams, flowing with water that came as clouds from across the seas then dropped as rain upon the land. All that lives is part of something beyond itself.

Each of us is like an actor in a play and no matter how small or unimportant we may at times imagine our role, it is an illusion to think we are not interconnected and interrelated to something magnificent. Everyday the awesome powers of nature, earth, fire, water, and wind circle the globe and reach into and out of the cells of all living things.

Earth and water are dissimilar, in reality opposites. One is a solid, the other a liquid. Nature often brings together dual energies that balance and compliment each other. For example, earth needs water to soften it. Without water earth becomes hard then cracks before turning to dust. Water's purpose is to feed and nurture. In return for water moistening soil to make it productive, earth blends with water, creating a firm foundation from rocks, soil and minerals to make what grows strong.

Fire and wind are another example of two entirely different entities that blend well. Fire needs air to burn or fire will die. In turn, wind needs the sun's heat to get it moving.

Men and women are different from each other, but without the attraction of opposites, life would be as boring and impossible as riding a teeter-totter alone.

Nature's extraordinary combinations of dissimilar elements are designed to work together rather than pull apart and create separation. Shakespeare wrote of nature's duality in his magical fantasy of love, *A Midsummer Night's Dream*. In the play, which mostly takes part in a forest, mayhem results when love is thrown off balance, just as chaos results when nature's balance is disrupted.

Nature repeatedly takes two entities and makes each unique, yet equal. One of the play's main characters is Oberon, the King of the Fairies. He symbolizes masculine energy in nature represented by the sun and wind. From Oberon's home in the forest, he commands the spirits of the earth, ruling with the fire of his personality and the force of his voice. While he can appear as terrifying as fierce winds and raging fires, underneath his forcefulness Oberon also exhibits the gentleness of calming winds and the sun's warmth.

Oberons's wife, Queen Titania, represents the moon and earth. A beautiful, proud moon maiden, she possesses a strong sense of independence. Living among the stars, she moves the tides and stirs the senses. Her power over the seas brings water to earth, which she does by working with Oberon's power over the sun and winds. Titania's exotic moonlight exists by reflecting Oberon's sun. In turn, Oberon would loose respect, if he abandoned the earth to the long dreary darkness that would exist without moonlight.

The sun and moon—the only ones of their kind that we see from earth with the naked eye—are a counterpart to the other. The sun and moon, which represent male and female energy, achieve balance in a way that makes neither less than nor more than the other. The sun and moon both have their moment to shine without challenge or attempt to dominate the other. During the year, the summer sun shines most of the day, and is visible far longer than the moon. Yet, when winter arrives the reverse happens since darkness comes early to provide more time for the moon to light the earth.

For millions of years, the sun, wind, water, and earth have maintained a beautiful equilibrium, especially in the tropical rain forests. The vast equatorial forests of Africa, Asia, and South America produce great amounts of oxygen. The sun along the equator produces much heat that evaporates the rain that falls causing moist clouds. Then tropical winds push the moist clouds across the seas over which the clouds pick up more moisture. When the clouds hit land, they drop life-giving water onto the continents.

Additionally, the forests are the lungs and cooling system of the planet, as important to the earth's function and health as the lungs and sweat glands of the human body are to our vitality.

Earth's oxygen comes from two main plants that have chlorophyll and use carbon dioxide to produce food. Those two make oxygen as a by-product. Those sources are green plants and trees and tiny one-celled phytoplankton that swim near the surface of the ocean. Earth would be uninhabitable without forests and phytoplankton to replace the massive amounts of carbon dioxide we exhale and produce. We need clean air that is rich in oxygen because out of the four elements—air, earth, fire, and water—we can live only a few minutes without the breath of life.

Tropical rainforests are different than those at northern latitudes. The tropical rainforests are far more easily destroyed. And, their destruction lasts a very long time. The fragility of tropical rainforests occurs because winters near the equator are so balmy that trees do not need to drop their leaves. Minimal leaf loss means that over thousands of years very little vegetation has fallen to the ground to make new soil. Consequently, the soil in the Amazon is quite shallow, only a few inches, and very poor. When tropical forests are cut, the earth loses the giant wall, or windbreak, that keeps the Amazon's limited topsoil from being blown away. Cutting the forests also removes the forest canopy, or umbrella, allowing rain to fall and wash away what little soil there is. When there isn't enough soil, seeds from trees have a hard time taking root. Due to the Amazon's lack of soil depth, *thousands of years* will pass before the Amazon's forests can return to the diversity and benefits they hold for the planet.

Tropical rain forests give the gift of cooling the planet and forming moist clouds that bring rain to the continents. If we take care of the trees, the trees will take care of us. But, if we cut down enough forests the earth's temperatures will rise, weather patterns on the continents will change. Phytoplankton in the oceans, being one-celled, can't survive but a few degrees change in ocean temperatures. Plankton is dying in record amounts. Without the oxygen plankton provides, fish can't survive. The changes in the planet have already started. As Puck says to Oberon in the play, "Lord, what fools these mortals be!"

The connection between earth and water is binding in a symbolic marriage similar to the sun and moon portrayed by Oberon and Titania. The earth's forests and the ocean's plankton are as necessary to our lives as male and female are to life continuing on the planet. Without one, there would be no other.

CHAPTER 38

EGGS IN THE NEST

I often watch the very over-active, yet diminutive little birds known as bushtits that gather in the coastal live oaks. Bushtits have a velvety-gray body with a brown crown. Barely reaching four inches from the tip of their beak to the end of their tail, bushtits are noticeably smaller than most backyard birds.

The little birds congregate in small bands, flitting energetically and continually from branch to branch. Thirty or forty of the busy birds keep probing and picking to glean tiny insects from the tree. They stay in constant touch with each other through a steady chitchat of soft chirps. The bushtits come and stay awhile in an area then they seemingly vanish. When they return, they take up residence less than an eighth of a mile away, but you may not know they are back if you don't see them.

From the bushtits enthusiastic nest building, it appears the flock intends to mate and raise their young in the oak grove. The nests resemble small hanging gourds made of twigs, soft plant wool and lichen.

From the top, the nests appeared round as are all nests. A mother bushtit keeps five to fifteen fertilized eggs at an ideal temperature with her body. Her feathers form a protective layer as warm as any down blanket. Inside each shell is the yolk, the future bird, surrounded by the egg white, the protective fluid.

Women recreate similarly. A human fetus forms from a fertilized egg. A pregnant woman's rounded uterus holds her developing fetus much like an eggshell holds the developing chick embryo. The uterus, akin to an eggshell, holds the amniotic fluid, the counterpart to the egg white.

Mother earth is also round and "gives birth" continuously. To safeguard the life mother earth brings forth, earth has a protective fluid, the seas, contained within a protective sac, the ozone layer. The ozone layer is proportionately no thicker than an egg shell.

Destroying the ozone layer is the same principle as poking a hole in an egg before it hatches. If the hole is large, the embryo will die because the fluids inside the shell will dry out. If a female uterus is punctured, the water breaks, the uterus collapses, and the mother is unable to support the life within. If the ozone layer gets too big a hole in it, mother earth's "shell" will crack more. Life as we know it will then be threatened further, and far more than bushtits will vanish.

Chapter 39

THE FOREST OF THE FUTURE

As a child I longed to ride a horse as fast as the wind. The nearest horse was a half-hour drive to a place that had an old gray speckled mare with a dull coat. Not the shiny black horse I longed to ride. The mare lumbered slowly around a small coral. She was so slow I could have fallen asleep in the saddle without falling off.

When I grew up, I decided if I was going to ride at full gallop, it was up to me to make it happen. So the kids and I went to a dude ranch in the High Sierras, north of Tahoe. The adventure was everything I imagined and more.

As I rode Ginger, a lively chestnut, the wind rustled through the trees. Taking a deep breath, I inhaled the sweet scent of pine-laced mountain air and thrilled when Ginger flew over huge fallen logs. The horse's incredible power further showed itself when Ginger climbed steep embankments and maintained her balance on steep narrow ridges as we traveled through the pristine forest.

All around us nature had woven a rich tapestry of textures that included the bristly white fir, the yellow-green needles of the ponderosas with their faint citrus smell, and the majestic, king of pine trees, the soaring sugar pines with their sweet sap. In sunny little meadows venerable black oaks grew and along the sides of streams we heard the soft rustle of the cottonwoods, alders and willows. When Ginger brushed against the lacey soft boughs of incense cedar the exotic scent filled the air. Sunshine danced between the trees letting light and shade play together on the ground. The unfamiliar songs of unknown birds filled the air, and beckoned us to respond to the forest's incredible rhythm. The forest seemed alive, an ancient, primordial, vital entity, and we were part of that aliveness.

Late in the ride, we emerged from the forest's shelter and entered a vast clear-cut. The forest immediately died.

Huge gray stumps stuck out of the ground like tombstones. Animals or birds no longer lived in the space where trees had once provided food and shelter and in return those same animals and birds planted the forest of tomorrow. I could taste the dry blowing dirt on my lips as the forest's coolness gave way to a hot, dry, sun-baked landscape scared by deep furrows. Snowmelt had carved the furrows since the sun melted the snow too fast. When spring brought rain, the trees normally held the rain in their roots, storing it in a watershed. But without the living trees, erosion muddied the river below.

"Used to be good fishing down there," the ranch hand remarked as we looked down at the river, "but the salmon don't come this way no more."

"How come?" my son asked.

"Salmon need clear water to lay their eggs."

"Do the trees ever grow back from the stumps?" a woman asked.

The cowboy shook his head, "Not ordinarily. The loggers leave short stumps only to slow the soil from being carried away," he paused, "works for about five years."

"How long ago was this area cleared?" I wondered.

He thought a moment. "Getting on ten years, as I recall."

"Why do we do it?" a teenager from Chicago asked.

The cowboy lifted the brim of his hat, looked him in the eye, "Cut-n-git is fast and cheap."

"I get that," he retorted, "but *why* do we do it?"

The trail guide kicked his horse to start to leave. "It's just the way it is."

"That's not a good enough reason," the teen said.

In the weeks that followed, I thought about the boy's remark. We've always had a weird relationship with our forests. We've loved them dearly as peaceful sanctuaries, as magnificent wilderness, and as a place where we feel free of city noise. Yet, we've also treated the forests dreadfully. In the nearly four hundred years since the colonists arrived, when old growth hardwood forests covered the continent, we've cut so many trees that only about three percent of the old growth remains. Even the old growth that remains is threatened. In temperate forests, trees can grow to marketable size within 50 years, but the layers of diversity that took centuries to develop are precious; once lost such diversity doesn't return in a lifetime.

As a result of public outcry, clear-cutting of public lands is down considerably in the United States, but it continues in other places.

In Wisconsin there's a certified sustainable, managed forest that has operated successfully since 1865. A multi-national corporation does not own it. The Menominee Tribe of Wisconsin owns and has logged the timber for going on 150 years. It is considered a model for forests in the new millennium.

The Menominee greatly limit clear-cutting of shade dependent trees to narrow sections rather than clear—cutting enormous tracts. The narrow clear-cuts leave enough trees so the forest appears a shadowy image of its former old growth self. The cuts are away from streams so erosion doesn't disturb fish habitat. The healthy, biologically diverse tract yields nearly 30 million board feet a year. In contrast to the Menominee logging practices, in the Western states whole sides of mountains have been clear-cut for decades. The practice left scars visible for miles.

The Menominee only allow heavy equipment that has rubber tires so as not to disturb the soil since vehicles with heavy tractor-type tires can damage the forest floor. Instead of clearing the forest floor to sell for pressed wood products, they leave boughs on the ground since boughs contain seeds. The boughs cover the cut area so the soil holds moisture to help the seeds sprout. To protect the sprouted seedlings and underbrush, the Menominee only allow dragging of trees on permanent trails.

The tribe steadfastly trains loggers before they can log in their forest. Without question the tribe will fine loggers for cutting unmarked trees or bringing other harm to the forest. Traditional logging practices cut the largest trees to get more lumber out of each log, but the Menominee do the opposite. They preserve the biggest trees so the genetically healthiest trees re-seed their forest.

Two different philosophies are represented in this story, traditional logging and Native American forestry. Maybe one day the roots of the two philosophies will blend with wisdom and understanding. One day the Environmental Protection Agency may listen to the wisdom of Native Americans in their forest management policies. Perhaps, a hundred years from today, my great, great grandchildren will ride a horse through the Sierra Mountains and not spot a barren clear-cut that is twenty miles away.

It could happen.

CHAPTER 40

A GIFT TO GIVE

In Southern California's semi-desert, the Chumash Indians did not rely on corn for their main food supply, but they cherished the oaks for providing an equally important food source. The only problem was the bitterness and toxicity of the acorns as they came off the trees made people sick. Eating raw acorns leads to painful cramps. However, Native Americans believed that everything from the Great Spirit had a gift to give. Their belief motivated them to find a way to use the thousands of pounds of acorns that mature oaks can yield during good years. It was also important to find a process for using acorns because they offered abundant nutritional benefits, including fat, protein, calcium, magnesium, phosphorus, potassium and folic acid.

Each August, the Chumash gathered enough acorns to last the winter. They set the acorns in gigantic baskets to dry in the California sun then stored them to use as needed. Weaving the huge baskets that were several feet across was as impressive as figuring out how to render acorns non-toxic.

To remove the bitter tannins, the Chumash used rocks to crush acorns to make a coarse meal. By increasing the surface area the task of leaching the tannins went faster. Leaching involved passing water over the acorn meal many times until the water turned from milky to clear, indicating the tannins had leached out. The Chumash made acorn soup, which they ate at almost every meal, by mixing acorn meal and water in a tightly woven basket into which a red-hot stone was placed and the mixture stirred until thickened. The acorns can also be made into flour. After cooking, leached acorns take on a slightly sweet taste.

Besides acorns, Native Americans figured out how to use thousands of items. The belief that everything has a gift to give encompassed plants, animals, trees, rivers, rocks, and people. Consequently, they approached nature with awe, curiosity, and excitement, like the way kids eagerly look forward to opening Christmas presents. Instead of presents from Santa, Native Americans wanted to discover what life supporting treasures a loving and wise Great Mystery had hidden for them in nature. They never doubted that there was a purpose, a gift hidden inside, and that everything on earth deserved their respect.

Their faith coupled with keen intelligence led to a cornucopia of discoveries. By the time the Old World met the New World, Native American's naturopathic knowledge was light years ahead of Europe's where leeches were still used. This fact soon became obvious to those that set foot on these shores.

When the famous French explorer Jacques Cartier navigated up the Saint Lawrence Seaway, winter set in early. Ice flows completely closed the river that year, so that making his way to the open sea became impossible. Cartier along with a hundred sailors had no choice but to spend the winter. They camped among the Huron, a large tribe of Native Americans who originally inhabited Ontario, Canada. The Huron, also known as the Wyandot, hunted in winter and farmed in the spring and summer.

Cartier noticed that, although some Huron got scurvy, they recovered within a week or so. The explorer had already buried a sizeable portion of his crew from the dreaded disease that causes hemorrhaging of body tissue, debilitating joint pain, bleeding of gums, loss of teeth, and bleeding around hair follicles. From the Huron's medicine man, Cartier obtained a concoction made principally of pine needles and bark. After eating the mixture, his men recovered.

A couple hundred years later, James Lind, a British naval doctor, read Cartier's account of the Huron's amazing cure. Western understanding of chemistry had advanced considerably by then. After the doctor figured out pine needles are acidic, he speculated a diet deficient in ascorbic acid caused scurvy. Lind experimented while at sea by feeding a few infected sailors oranges and lemons; the sailors quickly recovered. Impressed by the doctor's evidence the King of England ordered that throughout the empire all British ships had to carry lemons and limes. That's how the English got the nickname Limey's.

The Huron's use of pine needles to cure scurvy thus led to the discovery of Vitamin C and detection of other vitamins. The list of Native Americans' cures from the fields and forests of the Americas was long and sufficiently respected that doctors in the Old World knew Native Americans held the key to a very rich minefield of pharmacological knowledge. Word of Native American medicinal practices spread throughout Europe's learned. Humans are curious creatures, especially when lives are at stake. The vastness of Native Americans' wisdom prompted much scientific study. In the newly formed United States small, inexpensively printed books of Native American cures were compiled and widely sold. Pioneers living along the ever moving westward frontier often relied on those books, and on the intelligence of nearby tribes to identify and teach them how to use local plants.

From the forests of North America, Native Americans made remarkable medicines that are still used today. For hundreds of years, the bark of the hemlock tree provided an important ingredient in a tea that seldom failed to cure influenza. From a parasite growing on oak roots came blue cohosh, which Native Americans used to treat menstrual cramps and menopause. In the woodlands the wild trillium served to ease pain during childbirth, and pioneer women relied on the wild lily for that purpose. The inner sheath of a long, gray tree moss called Old Man's Beard contained an ingredient to help cure nail fungus. From the yucca plant came a foamy detergent and shampoo. Kelp from the sea, rich in iodine, helped prevent goiter. From the willow tree the People developed a liquid concoction for curing headaches that contained salicin, closely related to the active ingredient acetylsalicylic acid, better known to the world as aspirin.

Native Americans developed medicines for everything from intestinal disorders to seizures. Some of the more common items they gave the world include wintergreen, witch hazel, and petrolatum jelly.

Native Americans believed that everything the Great Creator put in the universe had a purpose, a gift to give for the betterment of the whole world. The People (as Native Americans often called themselves) understood that different parts within the whole plant worked together, the same way a tree's leaves work in harmony with its invisible roots. The People's cures also combined two different plants, enabling one plant to either boost the curative power in the other or to counteract side effects. The People didn't embrace a divide and conquer approach to medicine. Thus, they did not seek to isolate one ingredient from a plant, which can lead to toxicity, if taken over an extended period of time.

Besides medicines, Native Americans developed revolutionary surgical techniques that were a far cry from leeches. Throughout the Americas surgery was performed with razor sharp obsidian that was so sharp bleeding was minimal. Amputation was rare among Native Americans. Wounds were sutured with a strand of a patient's own hair. The wound didn't become infected since the person's own hair was not a foreign object that the body would reject. In the Andes, the People performed brain surgery, and developed syringes.

Long before psychiatry, the People saw a connection between mind, body, and spirit, which was consistent with a philosophy that saw all of nature connected in the web of life. Believing that the mind, body, and spirit were connected, the People did more than determine what was physically wrong, they asked "why" an illness occurred.

Native American mothers kept their children's placenta as a reminder of the child's connection to its birth mother and mother earth. In case the child became seriously ill, the dried placenta could help heal the child. We now know the placenta contains a person's own stem cells.

For all the healing gifts Native Americans provided, there was still a tremendous amount of knowledge lost. If only history had been different.

What happened to Native American culture was predicted long before Columbus sailed. On one of the three mesas in Arizona where the Hopi live, I had the privilege of seeing an ancient carving in the rocks. The carvings predicted the white man's coming and the choices that would effect the People. The carving consisted of three boats in a vertical column. Each boat had a different flag that symbolized three possible ways the new arrivals thought.

The first boat carried a flag with a long, narrow cross. The starkly linear symbol meant more than Christianity. It also stood for hierarchy or a group arranged by class, such as royalty, elitism, as well as a certain strictness and rigidity. The prophecy warned that, if the strangers arrived bearing only a cross, dire consequences were predicted for the People.

If the people in the boat raised a flag with a circle on it, life would remain unchanged because the stranger's basic beliefs were the same as Native Americans. Those are found in the eternal circle, which represents equality, a holistic approach, respect for the earth and creation.

If the strangers came with a flag that had a circle in the center of the cross, much like a Celtic cross and the early Christian cross, great advancements and happiness would grow out of respect for both philosophies because both had a gift to give to the other.

As we enter a new millennium, the once vast philosophical differences between Native Americans and Western beliefs are growing closer. We can see it in the environmental movement, the growth of naturopathic medicine, in a greater appreciation for diversity, and a growing admiration for Native Americans.

Throughout the world people seek to protect the earth, but on every continent at an alarming rate we are still losing ancient, incredibly diverse forests. Our knowledge is incomplete as to the medicinal cures we may destroy that will help cure illness not yet present. There are gifts from the Great Spirit that are as yet undiscovered. Having cut so much of the earth's forests during the last century, those alive today risk becoming like the Grinch that stole the unopened presents around the tree.

Just as the acorns from the oak trees have a gift to give, each of us has a gift to give. One of our greatest gifts to future generations will be to ensure that the gifts of nature remain intact for them.

CHAPTER 41

TOXIN GOBBLERS

In the late 1980's the National Aeronautics and Space Administration (NASA) began testing household plants to discover which would work best to reduce pollution inside future space stations. Scientists knew trees and plants clean the air and supply fresh oxygen to us, but wondered if plants would do the same in outer space. In long-term space stations, astronauts may need a breath of fresh air. Here on earth when a person is breathing recycled air all day in an enclosed office, the person can go outside at lunch and breathe deeply to clear one's head, but in space one can't just drop outside. As a result of testing plants, NASA scientists discovered some unexpected findings that we can apply to breathe easier at work and at home.

NASA's former Senior Environmental Research Scientist, Dr. Bill Wolverton, found that more than a plant's leaves clean the air. Powerful tiny microbes in the soil and roots ate air-borne toxins as if they were hungry, little Pac-Mans. Wolverton and his team set out to ascertain which plants gobbled specific toxins from the air. The tests were conducted by putting plants in sealed chambers then injecting the chambers with the three most abundant indoor toxins benzene, formaldehyde, and trichloroethylene.

Bill Wolverton has since retired from NASA, yet he

continues his research through his non-profit organization Plants for Clean Air Council (PCAC) and his company, Wolverton Environmental Services. PCAC and the EPA both report that benzene, formaldehyde, and trichloroethylene are found in the following:

> Benzene is a widely used industrial chemical. It is found in crude oil and is a major part of gasoline. Benzene is used to make plastics, resins, synthetic fibers, rubber lubricants, dyes, detergents, drugs and pesticides.
>
> Formaldehyde sources include cigarette smoke, plywood, particleboard and other pressed wood products. It is also used in manufacturing carpets, draperies, permanent press-clothes, paper goods, and adhesives. Although these toxins are widely used, their rate of releasing formaldehyde generally decreases as products age.
>
> Trichloroethylene is frequently used in the dry cleaning industry. Additionally, it is used in varnishes, paints, printing inks, and adhesives. It's the parent compound of DDT.

As you can tell from the above list, the three toxins are so prevalent that it's hard not to come in contact with them everyday to some degree. The products made from the chemicals can off-gas, releasing pollutants into the air.

When buildings are well insulated and tightly sealed, it can trap the pollutants inside. NASA's laboratory tests showed that in the firmly sealed "Biome" ordinary houseplants removed an amazing eighty-seven percent of the toxins within just twenty-four hours. Few of us live in a sealed chamber, but as mentioned earlier our homes and offices are becoming more airtight to conserve energy. In addition, the need to preserve forests to combat global warming has necessitated increasing use of particleboard and/or pressed wood products, the very materials that contain one or more of the above toxins. Pressed wood products used in construction carry an increased risk of releasing higher than acceptable levels of harmful toxins as today's modern homes age if—*and this is a very big if*—building materials are improperly installed or homeowners neglect upkeep. Moisture penetration from leaky roofs, windows or plumbing rapidly weakens composite wood products. This means it is essential to protect composite woods as manufacturers direct, with several layers of paint and by preserving roofs and flashing on homes to prevent moisture penetration. It is also important to keep moisture from getting underneath tiles in kitchen and bathroom cabinets made with pressboard. The above precautions will not only protect one's investment, but also provide health benefits.

That said, the good news is that houseplants are nature's cheapest, most proficient, low-tech air cleaners. Plant leaves, roots, and soil absorb benzene, formaldehyde, and trichloroethylene. The Plants for Clean Air Council recommends combating air pollution by using two houseplants for every hundred square feet of floor space.

The PCAC advises on their web site that although *every* leafy indoor plant and many flowering plants help combat indoor toxins, certain plants are invaluable for specific pollutants. For example, the super easy to grow philodendron is ideal near pressed wood products. If someone in your house smokes, fill the house with a preponderance of English ivy. Old Victorian favorites such as the Boston fern and orchids are good to use around items containing formaldehyde and around adhesives. If you work in a clothing store with lots of synthetics or have tons of clothes choose the spider plant and Janet Craig dracaena. If someone in your family works in a gas station, Marginata dracaena can help remove gasoline toxins. For new carpeting smells buy a couple golden pothos. If you work around printers ink or plastics, chrysanthemums are good for removing airborne toxins.

It's amazing to consider that a little six inch potted houseplant can do for the air in our homes and offices the same thing that trees do for our home planet. Everything living does have a gift to give.

Chapter 42

BALANCE

While hiking in nearby Malibu Canyon, I noticed that an oak tree was growing out of a narrow opening in the rocks above the dam. The steep sides of the canyon put the oak in a precarious position since it grew out of the rocks at an awkward angle. If its trunk didn't straighten out, gravity would take its toll. I wondered if the boulders around it would stunt the oak's growth, or would the oak dislodge the boulders. A lopsided oak was mighty strange to see. Oaks normally have an uncanny knack for standing up straight, and all that good stuff; they are like the kid in school that seemed to naturally do a lot of things right. The lopsided oak tree got me thinking about the importance of balance.

I think that we are, like most oak trees, born with an innate sense that balance is essential to our well being. If our trunk, or spine, is out of alignment, it can change the way we walk and cause serious pain.

Far back in antiquity, humankind understood how meaningful and vital it is to maintain balance, not only within the body, but also for the earth. In time, a pair of scales became a symbol signifying the importance of the universe's sacred geometry of balance and design.

Coast live oak above dam, Malibu Canyon, CA

The idea of a woman holding a pair of scales came down to us through Greek and Roman mythology. The familiar goddess, known as Justice, wears a long white Grecian dress. In her right hand, she holds up a set of scales while her left hand holds a sheathed knife that points downward. Her responsibility is to protect nature's laws so that earth, wind, air, and fire stay in harmony and balance.

In time, Justice came to symbolize not only nature's laws, but also the laws governments make. It was during the French Revolution that Justice received a blindfold when the French decided their courts were prejudiced and not impartial. Both versions of Justice—with and without a blindfold—are in government centers around the world. If Justice is depicted with a blindfold or without one, her responsibility does not change. She must keep the scales balanced so they don't tip too far in one direction. To fulfill her ancient responsibility she must weigh both sides of an issue then make an impartial decision to protect the laws on which life is founded.

The ancient symbol of scales, relating to nature's laws, is a clever pictogram because the scales move. They are not fixed with super glue. Nature's laws are extremely flexible so that life on earth can continue by providing a balance between order and disorder. In other words, nature, including us and the choices we make, are a balancing act that reconciles the inevitable chaos resulting from freedom to change, evolve, make mistakes and recycle. Disorder is balanced against the orderly laws of physics, mathematics and codes in DNA, which provide organization and structure to all that fills our world.

The moveable scales correspond to the right of every creature to adjust in order to fulfill its purpose of giving its gift for the betterment of the whole. This is important because when the earth is balanced, it becomes productive, efficient and healthy; but more than that, it is beautiful, creative and able to heal quickly. When we are balanced, we are much the same. We are healthy, optimistic, and able to roll with life's changes and accomplish much for others and ourselves. When the earth gets seriously out of balance, nature's laws demand corrective action to protect and preserve life on earth. In the same way, corrective measures are needed when our bodies or lives get out of kilter.

Oak along Highway 101 near Geyserville, CA

Justice often appears blind to individual and local hardship because her jurisdiction is global involving laws that apply to the entire planet. When there's global warming it can cause droughts, deadly searing temperatures, forest fire conditions, and severe hurricanes. The goddess must take care of mother earth and adjust the flexible scales to bring the earth back into alignment. She has to let the great law of freedom function and that sets in motion the law of cause and effect. Our choices led to global warming. Justice stays focused on her duty, which is to re-adjust the scales by doing what is needed to stop what is making mother earth sick.

There are many laws of nature, but there are a few in particular that come to mind. The first is that the sun's rays warm objects they touch. We know it's cooler in the shade than in the sun. Cut down the forests and there's less shade, causing more sun to hit the earth.

Second, heat increases molecular activity. It may sound technical, but it's simple. Think of cooking pasta and imagine that the water in the pot represents the ocean. As water heats, it starts to move around, bubbles form and before long hot air rises. Often you feel the hot, steamy air moving against your face. Just as turning on the gas under the pasta pot sets the water molecules in motion, heat the oceans with global warming by only a few degrees and widespread changes take place. When warmer seas warm the air, the wind increases, resulting in more intense storms over a wider area of the globe. When the oceans are too warm, we get El Niño conditions. Heat the oceans over nature's "fudge factor" and the microscopic phytoplankton finds the water is too hot and it struggles to live. A lack of phytoplankton depletes the ocean's oxygen, and if severe enough, fish die. If the oceans continue to warm, maintaining balance might one day require melting the icebergs. This is no different than putting an ice cube in a cup of hot tea to cool it to avoid burning one's tongue.

When mother nature is out of balance, she acts weird, and in that sense she's no different than any of us who get out of sorts. Nature's laws require balance be maintained by whatever means so the beauty, health and wonders of the earth can work in harmony. We are—for the most part—civilized, law-abiding, well-intentioned people, who often unwittingly do dumb things that violate nature's laws. I'm not a gambler, but if I were, I'd bet on the goddess holding the scales. Nature always wins.

Chapter 43

A NICE NEIGHBORHOOD

Several hundred years ago, a squirrel stashed a little acorn in the tall grasses growing in a meadow surrounded by the walls of a steep canyon. When it rained, water flowed down the sides of the sandstone cliffs toward the lowest point in the canyon. A stream formed that became known as Malibu Creek. As rainwater seeped downhill, it moved past the acorn, moistening the ground around where the acorn lay. The water that flowed to the creek continued to meander for several miles before tumbling down to sea level after which it flowed into the Pacific Ocean.

Some inner knowing informed the little acorn that it had come to rest in a highly desirable neighborhood. Indeed, it was a lovely spot to establish itself. Much like when we find the right house and know immediately it is where we want to live.

The oak grew rapidly its first years, but it had to reach twenty years old and "grow up" before it could make acorns that enabled it to reproduce; much like we encourage teens to finish high school before taking on adult responsibilities. Nature exemplifies the same wisdom.

Every century since the acorn gained a foothold in the canyon, it reacted to both mild and severe changes in the environment. When it snowed or Santa Ana winds caused blazes that swept across the fire-prone land, the oak's thick bark protected it. Like us, the oak already had in its makeup everything necessary to meet the challenges that came its way.

Coast live oaks, Malibu Canyon, CA

There arrived a time when the area's temperatures soared, drying out the soil. When the creek became a trickle, the oak proved it was a fine neighbor. The oak's deep roots drew up water from deep underground pools. The moisture was distributed to the oaks surface root hairs and shared so that sage and other native plants could survive the drought. Despite the stress that the area was under from the drought, giving up was not an option. The oak curled and thickened its leaves with a thin waxy covering to lessen evaporation and preserve the liquid it had until conditions changed. In much the same way, we conserve our liquid assets and reduce expenses when money is in short supply.

Another year the rains came in torrents and the creek rose ten feet. The oak got soaked plenty, nearly drowned and could have been washed away. But the oak managed to hang on against the rushing creek because it had wrapped its roots around the boulders buried deep in the ground. In a similar way we hang on by our roots, relying on family and friends who are "rocks" on whom we can depend.

After the rushing waters stopped, the creek flowed gently again for many years. Life returned to normal. Hard times never last. Continual change is one of the laws of the universe. Each season for at least the last three hundred years, the oak met the challenges the weather presented. It even endured earthquakes and gale force winds. Now the oak stands magnificently as a reminder that hard times are part of every life. The oak did not ignore what was happening. It did not oppose the rushing water's flow, or resist the drought, or refuse to go along for the ride when the earth shook. Faced with unanticipated and undesired changes, the oak made changes. It went with the "flow" seeing in its canyon location that life is really a series of hills and valleys, highs and lows. By not fighting against the changes that it couldn't control, it adapted when the weather disrupted its tranquil world, and it survived. It never acted as if it was powerless because indeed it was not. It was innately resilient.

The self-reliant oak determined when to shed its leaves, the best day in spring to bud, and the ripeness of the acorns before it dropped them. It even determined where and how far to spread its branches without elbowing neighboring oaks. The freedom the oak had to decide to do all these things was innate and personal. Although other trees surrounded the oak, as other people surround us, the oak stood on its own in the meadow and could not forfeit its right to self-determination, nor can we.

The self-reliant oak does not function with an "I can't approach" attitude. It's we humans who tend to limit ourselves with fears of failure, rejection, or unworthiness. As the old tapes play in our heads, we impose limitations on ourselves. Plants and animals don't hear the thousands of admonitions humans hear in a lifetime. Animals are lucky in the sense that they make no prisons of their own design. Nature seems to implore us to return to a natural state where we say to our dreams, "I can, I can." Then, like the oak, ask to receive, trusting in a loving universe.

Sometimes it seems we've lost touch with what it means to be wild and free. In nature, there is a state of freedom where instincts are trusted and happiness is pursued, in the sense that actions are kind and loving to the self and others. Animals tend not to engage in self-destructive behavior. They demonstrate a regard for themselves and are true to what they are. All of nature—from the tiniest insect to the biggest elephant—is true to what it is. Nothing in nature tries to be what it is not; in the inherent truth that is manifested, we see beauty.

All wild things automatically rely on the earth for support and in return act with respect for it. Humans are not so fortunate. We have inadvertently and without intention jeopardized our soil, water and air. As the birds rebuild their nest, we will do the same for the earth.

Coast live oaks in Malibu Canyon

The oak has been a fine neighbor to all that lived in the vicinity. The canyon is similar to the type of neighborhood where everybody knows everyone else and looks out for each other. The oak was not inconsiderate. It took only what it needed from the earth leaving what it didn't need for others. All its life the tree gave back so as not to limit the liberty of other creatures living nearby. Although the oak took care of itself, it always affirmed its connection to other creatures in the web of life. In turn the birds, animals and all that breathed returned something vital to the oak. They gave the oak carbon dioxide that the oak turned into food with the help of sunlight. In return the oak gave back oxygen in a mutually beneficial relationship between green plants and creatures that must breathe the air. Because what is given generously is returned, the oak flourished.

The oak gave safe shelter and food for many creatures, much like ourselves when we invite family and friends to stay for a visit. Mother deer came to nibble the oak's acorns and with such rich nourishment the deer made milk that provided rapid growth for her fawns. Little lizards ate the bugs crawling on the oak's bark to keep the tree from becoming infected. Birds and squirrels shared the oak's acorns just as neighbors share flowers and vegetables from their gardens.

Over time, the oak grew bigger and more beautiful. When it had lived to a ripe old age, the acorns the oak had given as a gift had multiplied tenfold the same as the kindness we extend multiplies in ways we may not always see. From the oak's acorns new oaks were born and grew on the mountainsides and near the creek where the birds and squirrels planted them. In time, a vast oak forest formed that was like a family that stays close to each other and welcomes friends to join them. Indeed, without any cash expenditure, the oak forest resembled a bunch of friendly neighbors that made the area a pleasant place for all the other creatures that dwelt in the canyon. Life there resounds with beauty, joy, and trust.

Chapter 44

NATURE SHOWS THE WAY

It's funny how many things you can learn from being outdoors in nature. When standing in the oak grove, I've had glimpses about life and love that only lasted a split-second, but the insights provoked much thought. Sometimes I spent days turning the ideas around in my mind, examining their depth from many angles.

Nature opens up a whole realm of learning possibilities for everyone because some part of nature speaks to each of us. Before people turned to books for knowledge, nature was the world's teacher. This seems appropriate because it's unreasonable that a loving God would ever put humankind on earth for eons without giving clues about how to live in peace, health and happiness. I often wonder what we could have learned, if nature had not been seen as something we had a right to do with as we wished, but as a teacher with beauty and loving wisdom to share.

Wisdom is found in the principles nature adheres to each day. They are the same maxims many parents teach their children: speak the truth, clean up your messes, and play fair to give everyone a turn.

Speak the Truth.
The natural world is so inherently truthful that innumerable fields of science from astronomy to zoology exist

because we can acquire accurate, scientific information from nature. One field, dendrochronology, the study of tree rings, reveals a tree's life—past droughts, fires, floods, a tree's own DNA, insect infestations, and even global weather—as surely as we record our life's events in a journal.

If a forest fire burns thousands of acres of timber to the ground, we can still learn the fire's cause from the ashes because nature keeps an accurate record. Nature does not lie. Whatever it does there's no need for denial because nature is naked as a newborn. The natural world is fully able to withstand the closest inspection. The information exists for anyone to see. Nature's workings—more complex than the largest conglomerate—do not require spin-doctors or cover-up detrimental conditions.

Years ago, I saw an old man of such dignity and humility on Oprah Winfrey's show that I've never forgotten him. The topic was contaminated ground water in Louisiana. Many lawyers and company representatives filled the audience and affirmed they'd broken no laws. They'd complied with all state and federal laws, but didn't mention the loopholes by which they'd operated.

The old gray-haired man in the audience was slightly stooped and dressed in worn clothes. He stood up and supported himself by holding on to the chair in front of him. He said that he didn't know about all those laws they were talking about 'cause he wasn't an expert on laws. He only knew that his dog went into that company pond and when he came out, his dog had no hair left. His bald dog died the next day. Then the man sat down without a look of triumph. He merely stated his truth. He reminded me of nature the way he said it. Nature is not about blame, guilt and punishment. It's about being visible to all.

If you make a mess, clean it up. Don't leave it for somebody else.

Since everything nature creates is biodegradable, it makes environmental cleanup extremely proficient.

Nature's maid service is free, dependable, high-quality help. In the process of getting rid of discards, nature actually helps the planet. A mountain of autumn leaves turns into rich loamy soil. Animal waste becomes fertilizer. Boulders eventually turn to sand on the beach. Imagine millions of years of no such maid service. We'd have long ago suffocated in nature's waste, if nature was designed to be any less helpful and supportive. Nature's cleanup system is incredibly efficient, ingenious and considerate to all that dwell upon the earth. Nature creates no toxic landfills or hazardous waste disposal facilities. Perhaps that's what the saying "cleanliness is next to Godliness" is really all about.

Play fair. Let everyone have his or her turn.
Nature places much responsibility on the local neighborhood to work together to support that which creates and promotes all life. For an understanding of how this works, pick any area of woods across North America, perhaps woods beside a highway you regularly travel.

The woods add beauty, help clean the air, and so much more. In winter the woods form a living fence that can hold snowdrifts and divert strong winds. The trees have thousands of roots that aerate the ground. When spring rains come the root hairs absorb water to help prevent flooding. In summer, the same roots pull up moisture from underground streams enabling plants with shorter roots to get water so they'll survive. Summer shade cools the woodland protecting herbs that heal us and wild flowers that bring delight. Over the cold winter, autumn leaves make a warm blanket protecting plants when temperatures drop below freezing.

Imagine, in the same woods, a fierce windstorm snaps a very old oak in half. As the thick trunk breaks, it sends a deafening crack through the air seconds before it falls onto the ground. The distinct sound is far louder, yet similar to the crack of a baseball bat breaking. When the tree crashes, the ground under your feet vibrates. Although the oak no longer "works" in an upright position, it continues to benefit the woodland.

For the next hundred years, more or less, depending on the climate and size of the tree, the downed oak will give many gifts to the woodland. During the century when it appears to lay down on the job, many local creatures come to the log. Events take place that affirm that every life, even the tiniest insects, have a purpose in God's plan. If a little bug is important, as you will see it is, how can anyone doubt that he or she is not equally valued? Or, that each of us doesn't have a purpose.

A host of local creatures depend on the log for food, water, shelter and "employment" the same as a local company employs and supports a community. The following are a few of the innumerable players that will work at the log, coming to it in an orderly manner and sequence.

Before the tree fell, members of the large and varied woodpecker family drilled rows of holes in the bark after the oak started to die, but before it fell. A male woodpecker may have chiseled out a niche for a nest. After the tree fell, niches and holes serve as tiny bowls, collecting rain to aid the log's decomposition and provide water for insect survival in dry weather. (How many times has each of us performed a random everyday act, like a woodpecker doing it's job, and felt good later when we found out the positive impact the act had for someone else?)

Spring rain and wind carry spores from mildew, mold and fungus that collect in the moist holes and start working on the log, preparing the way for various insects to do their job. Brown rot fungus produces enzymes that feed on the cellulose, which is in all plants. A noticeable stringy fungus, which looks like white sewing thread, later meanders along the length of the fallen tree. The white fungus broadens the spaces in the wood grain so moisture gets in and makes the wood spongy. Moist green mosses soon cover the log.

The mosses soften the thick hard bark by keeping the drying property of air off the trunk.

Thousands of different beetles exist. Many specialize on a particular tree species, such as the oak bark beetles. They are lured to the log by pheromones given off by the damp log. We can't smell the aroma, but for several insects the smell is intoxicating. It causes oak bark beetles to treat the log as a romantic resort hotel. They make themselves comfortable between the bark and hardwood. It is there they mate and eggs are laid. When hordes of eggs hatch hungry little larvae ravenously snack away at the underside of the bark, creating a space that allows moisture and other insects to get in to further loosen the bark from the trunk. If the bark beetles get too prolific, their natural enemy woodpeckers and wasps will help keep the bark beetles in check.

More bark falls off thanks to chipmunks, squirrels and raccoons playfully jumping on the log. Hungry crows, with beaks longer and tougher than sparrows, peck at tasty insects hiding in the bark and in the process toss bark bits off the tree.

A different beetle, the wood boring beetle, has a special task of carving tunnels. These beetles get to the heartwood. When they leave, instead of leaving the way they entered, the wood boring beetles drill one-eighth inch tunnels and tunnel their way out of the log. When they leave, the log looks like it's been hit with buckshot. Rain and snow enter through the tunnels, providing, moisture the tree can't get since it's severed from its roots. The insects with their lack of dentures need moisture-softened wood to feed on cellulose. Hosts of wood destroying insects live in the forest. Unfortunately, we lure them to our homes by the smell of rotting wood from dripping pipes, a leaky roof or moisture in a crawl space. Insects, like us, lack chlorophyll to make their own food and require food and water the same as we do.

Nursing log, Tyron Creek State Park, Dunthorpe, OR

When the air warms, carpenter ants and termites get a whiff of the pheromones from the rotting wood. In spring both insects have wings and can fly through the air in search of a hospitable location to start a new colony. The colony eventually grows to include several thousand hungry members. Biological commands prompt the black, super-speedy carpenter ants to shred wood into skinny-mini toothpicks they toss out to create a hollow for their nest. Termites turn wood to powder so that toothless insects, such as sow bugs, worms, and regular ants, can do their job.

Eventually, fifteen to twenty years later, depending on the climate, a portion of the huge log rots away. The hollow offers a home for rabbits, a shelter for mice to hide from hawks, a cool repose for snakes and a nesting place for ducks. After forty to sixty years, more of the log deteriorates and then part of the log may resemble a hollowed out canoe. When surrounding trees drop their leaves and line the "canoe" a bed of leaves turns to rich compost over the winter. The next year, seeds from overhead trees drop onto the damp compost and take root in the loamy soil. At this point the old fallen oak has turned into a *nursing log* for baby seedlings that will become tall trees in the future.

The fallen log remains beneficial for decades. The portion of the log resting on the ground absorbs and retains considerable moisture during the rainy season to help the seedlings survive. The more decay between the spongy layers of wood, the more moisture the log absorbs. The moisture on it's underside provides a cool place for insects to withstand hot dry summers, much like bugs gravitate to the cool damp area under a flowerpot. Insects depend on the moist, coolness of nursing logs to insure their survival during a drought or a hot spell. The mother logs ensure that insects live on to recycle the deadwood and keep the forest healthy. In case there's an imbalance of insects, birds will fly in by the dozens, aware on some primal level of the need for them to eat the excess bugs and balance the forest's insect population.

Some may call the tree dead, and proclaim dead trees useless lying on the forest floor, forgetting that huge logs, forty to eighty feet long, occupy the space that would normally be taken by fire prone underbrush. In nature's Divine Plan the end is always the beginning of new life. A forest littered with big, moisture providing nursing logs promotes the life of the forest for years to come. In nature we can find the multitude of ways that nature takes death and transforms it, infusing it with new life, and no less happens to us when we die.

Nature's laws aren't written in a book, but the guidelines speak to us in silence, when we listen with our hearts. Then we find that nature urges us to passionately value the truth. It shows us the benefits to life itself when we create products that won't harm the earth and recycle so that all life is promoted and enhanced. The guiding principles originated from an immense love by nature's God for all things great and small. Our planet was designed to give every creature that flies, swims, crawls, or walks upon the earth surroundings that encourages each to give their gift by playing a personal role in the Divine Plan.

Chapter 45

THE DROUGHT

In Southern California's semi-desert, the oaks that line the coastal canyons depend on fifteen to eighteen inches of rainfall per year. The rains get absorbed into the soil then trickle through sandstone into underground streams. The giant oak's dependence on water makes them very selective about where they grow.

Without water the root hairs, like nerve endings, cry out for water to feed the tree, the same way our brains become fatigued without water. Trees die without water. We, the other Standing Ones, can't live without water either. A person can starve to death in a matter of months, but the same person will die in a matter of days without water. This chapter and the next are about the hardship caused when the rainfall dropped dramatically in Southern California.

All my life I took water for granted. I turned on the faucet and water came out. It was a given until the drought started in Southern California in 1987. The cloudless skies marked the beginning of a five-year drought, and the start of changing global weather, though few knew it at the time. The drought would stress the oaks and change bathroom habits across the country, proving that on earth the web of life is amazingly interconnected. What happens on one side of the web sends vibrations across the web onto the other side, much like a spider's web functions. But, that's getting ahead of the story.

During the drought's first few years, the normally golden grasses on the hills around Los Angeles turned a morose charcoal color that showed no life in the brittle stems. In movies filmed in Southern California during the drought, you can see the hills in the background appear black rather than golden. The price of water rose dramatically. The city asked restaurants not to serve water unless a customer requested it. Boat docks reached out like empty hands into dry lakebeds. Lawns died. Golf courses and car washes had to install expensive gray-water recycling systems to stay in business.

The lack of water caused bobcats, coyote, and deer to come down from the hills. One day, while stopped at a traffic light on Thousand Oaks Boulevard, I watched two coyotes use the crosswalk and then continue on their way to a McDonalds's dumpster in search of refreshments.

In an effort to explain the weather, the words, El Niño, indicating sea heat, and La Niña, indicating sea cold, entered the lexicon. For the next five years, La Niña conditions sent the rain elsewhere, which was highly unusual. In the 1880's—before the heavy use of fossil fuels—El Niño conditions happened once every ten to fifteen years. Five consecutive years of La Niña drought was unprecedented. As fossil fuel use increased, coupled with the cutting of more forests, a disproportionate amount of carbon gases collected near the earth's surface. Cutting forests releases tons of carbon dioxide that trees store in their roots, the same way we stock up on food to have for emergencies. An abundance of pollution and carbon dioxide formed an invisible barrier, much like a cover on a stockpot prevents heat from escaping. The ground-level heated air spread out over the oceans, making the oceans warmer. As the air currents and jet streams changed, so, too, did the weather.

In Los Angeles County, a lack of rainfall is serious because millions of people—roughly a quarter of the entire state's population—live there. Those who've seen the movie *Chinatown* may recall the intriguing story of the county's founding fathers' determination to "import" water down to Los Angeles by gravity feed. The water came from Mono Lake, located on the eastern edge of Yosemite National Park, 350 miles away. Mono Lake sits like a blue oasis 6,300 feet above sea level. Having available water for drinking and irrigation made Los Angeles real estate incredibly valuable, which was the plan. After decades of siphoning off fresh water from streams that fed into Mono Lake, the lake's existence was threatened and further implications loomed.

Millions of insect-eating songbirds that migrate between North and South America depend on Mono Lake as a stopover on the flyway. In the Pacific bird migration path, larger birds—geese, ducks, and gross—take the coast route where breezes are stronger off the ocean, which is considerate of small birds. Songbirds take the inland route protected from severe winds by a wall of mountains, the High Sierras and the Cascades.

Where the Sierras and Cascades end on their eastern slopes, the desert begins because rain clouds off the Pacific are too heavy to scale the high peaks that tower over ten thousand feet. The clouds drop most of their moisture on the western slopes of the mountains. That is why desert or semi-desert conditions extend from Arizona north all the way to part of the state of Washington. Since small birds take the less windy route on the edge of the desert, water is critical to their survival. And, songbirds are critical to controlling insects since they gobble zillions of insects.

After songbirds cross the Sonoran Desert their first main stop for water is the Colorado River. Then they fly across the Mojave Desert to the next watering hole, the fresh streams that feed Mono Lake. Having expended so much energy in migration the birds are thin and need high-energy food. Every year, about the time the birds arrive at Mono

Lake, a special fly larva hatches that's extremely rich in fat and protein. It's all part of nature's elaborate plan.

Half a century of sending water to Los Angeles put Mono Lake in jeopardy. The drought only made matters worse. When Los Angeles needed Mono Lake's water more than ever, the people of Mono Lake filed suit to save the lake. They won. A judge turned the tap to Los Angeles' water supply down to a trickle.

In Los Angeles, the lack of water became so severe that the city instituted strict water rationing, which affected millions in unexpected ways. To conserve more water the city issued guidelines for taking a shower: wet down, turn off the tap, lather, and then turn on the water to quickly rinse. That didn't go over big. Then the suggestion came for people to shower together, which brought few complaints, except water use didn't go down at all.

Then the city suggested not flushing millions of toilets unless necessary. Many found that disgusting. It wasn't long before the infamous low-flow showerhead and toilet were invented. To promote their use, the Department of Water and Power gave away free toilets. When free toilets met with less enthusiasm than expected, the city passed an ordinance requiring all new construction to install low-flow toilets.

Municipalities across the country adopted the new toilets as an advantageous way to extend their water supply while their cities grew. The toilets soon became the standard across the country.

Even though the water shortage forced Los Angeles to become a model of water conservation, it didn't solve the state's problems. Agriculture has first claim to 80% of California's water. The nation gets a third of its produce from California. Agriculture's needs caused the state to take the water held in dams, intended for the rivers, and divert that water onto farmlands. The resulting shallow streams and dry wetlands killed countless fish, birds and wildlife. When still more water was needed for crops, the state gave farmers permission to dig wells. The wells took the water from the water table down to, and sometimes including, the sediments that contained pesticides. The dangerously shallow underground streams stressed the oak trees that depend on that water. Worst of all, after five years of hardly any rainfall, California was on the brink of *desertification*.

Thousands of trees died. Others were so weakened by drought that their resistance was down and they became infected by the bark beetle. Nature's plan protects the forests from the wood-boring insects with deep, wet mountain snow packs. Without rain, the snowfall was very low in the Sierra Mountains. When there isn't much snow, the beetle thrives, and goes to stressed trees for moisture. Throughout the West, abnormally dry conditions sparked flames that burned 1,500,000 acres of forest.

California and the rest of the United States were not alone in the effects of La Niña. It was global in scope. The dry conditions caused ten percent of the Amazon rain forest to catch fire; ships had to lessen their cargo to get through the Panama Canal; coral reefs suffered when the water warmed; Malaysian forests burned; icebergs broke off; and Australia and East Africa suffered severe drought.

Finally, in January of 1992, way out in the Pacific, halfway around the world, a storm brewed. It rained at last. The rain started slowly at first with gentle drops that barely made a sound. Then the tempo increased with the blustery wind. The storm turned violent. Water slid off the rock-hard, dried-out hillsides like water runs off marble until enough soaked into the ground to cause mudslides. After only a few days with its fury spent, the rain sounded tired. Rain tapped at the windows playfully, gently, until it fell as soft as a kiss.

Soon after the rain stopped, the landscape changed dramatically. Nimble little grasses grew, replacing black hillsides with the look of apple-green velvet. Fruit trees flowered in shades of pink and white, scenting the air with

Oaks along Highway 101 near Geyserville, CA

sweetness. Acorns that waited patiently for years in fertile soil for the chance to come alive did just that. An amazing 135 tiny saplings soon surrounded an ancient oak, (the same oak where the little girl patted the tree on her way to school). During this time, the cycles of life, death, transformation followed by rebirth were everywhere. Nurturing rain turned the landscape into a veritable celebration of life.

By the time we celebrated a new millennium, scientists affirmed the earth was warming. The changes that will bring in the years ahead for the earth's weather offer incredible global challenges.

Yet, like the 135 oak saplings that re-affirmed the ability of life to renew itself after the rains came, there's much hope for the future. Water, like love, must flow. A place without water is like a life that knows no love. It is barren, dry, and harsh . . . in short, a desert. When there's normal rainfall, nature reflects peace and harmony. That same potential to create peace and harmony is also within us. It's always been there, just as it was within the 135 acorns waiting for water before they could grow. It begins within us when we feel so connected to the earth that harming it is viewed as unloving to ourselves.

Chapter 46

ALL THINGS CONNECT

The facts you're about to read relate perhaps the greatest lesson from the oak trees: what we do to the earth, we do ultimately to ourselves.

The drought that put California on the brink of turning into a desert affected me deeply. After five years of it, I began dreaming of green lush vistas and the constant music of water tumbling over rocks. I went in search of such a landscape. I pulled up my roots. Without knowing anyone, I moved to the Pacific Northwest. Now I live in a little house only two hundred feet from a large woods filled with towering maple trees, ash, holly, and cedar. The woods has two springs of which one them meanders and tumbles over moss laden rocks before it runs through the backyard.

I was only in the Northwest a few years when I started dreaming about California's beautiful twisted oaks that survived in large measure because their branches grow too crooked to offer much commercial value. The oaks gnarly, picturesque shape contributed greatly to their longevity.

The trees are so loved by Californians that they often add ten percent to the value of a home. Not everyone can have an oak tree since the oaks only grow where they can tap into underground streams. Oaks may grace one hillside and leave the neighboring ones bare except for grasses. Oaks grow slowly and live for centuries. In my former home, from my bedroom window I had the joy of looking at an oak that was over 600 years old.

When I moved to the Pacific Northwest, I brought with me pots of large jade trees that I'd planted as small clippings soon after each child was born. For several years the jades grew outside on the patio in California. After I moved north, a tiny live oak popped up in one of the jade plants. It was a wonderful surprise to find a California live oak had come along with me to the Northwest. It was a gift from one of the scrub jays or squirrels that compulsively plant acorns. It's been six years and the oak tree now stands nearly eight feet tall. I treasure that oak. Unfortunately, I recently received heartbreaking news that makes me treasure it even more.

Literally thousands of California's oaks are dying. It's a sudden death in which centuries old trees die in a few weeks. Death comes from a fungus related to the blight that caused Ireland's Potato Famine. Some scientists speculate the fungus was here all along, hiding, waiting for the oaks to face stressful conditions then the virus woke from its slumber and planted the kiss of death on the oaks.

Scientists attribute the oaks' plight to the use of fossil fuels that led to global warming. Like a falling stack of dominos, global warming caused weather changes that led to five straight years of drought, which stressed the oaks that had lived in California for centuries. The dry, hot La Niña years were followed by abnormally wetter El Niño years as the earth sought balance. The fungus thrives in water droplets, which is why it devastated the lush country of Ireland. The lethal spores attack the trunk's bark and branches producing an extremely toxic enzyme that dissolves the dead outer bark along with the living inner layers of the tree. Oozing red sores on the trees are the first clue since the fungus isn't visible with the naked eye.

Under a microscope the fungus resembles strands of cotton. By the time the tree is oozing, it's so weak that the decaying wood draws bark beetles. The bark beetles finish off the trees, contributing to the tree's sudden death, much like a weakened cancer patient dies not of cancer, but of say pneumonia. The great, magnificent oak trees that have lived for hundreds of years are disappearing almost in the blink of an eye.

The fungus is so pervasive that anything that comes in contact with the spores can spread the disease. Hikers are asked to clean their shoes, cyclists are asked to wash their tires, as well as the treads on cars. The dead wood is not burned. Everything and anything that contains the fungus must become quarantined.

The voracious fungus has caused the death of California live oaks, tan oaks, and black oaks. The microbe is so deadly that scientists' fear that it will spread to oak forests across the nation. It has already spread over 10,000 acres from Napa Valley to San Francisco to Carmel to Santa Barbara. Municipal water districts are fearful because trees that have live for hundreds of years are disappearing in a matter of weeks.

Without thousands of oaks on the hillsides, there will be nothing to absorb the heavy rainfall. The oak's roots hold the water then release that water slowly through the roots. The condition will deteriorate watersheds and cause higher concentrations of farm chemicals to spill into reservoirs.

That's not the only fallout. Trees are great rooted beings. They're fixed in the ground. They can't fly away to escape from a fire or migrate when a fungus attacks. Acorns form the framework for California's life support system for an estimated 150 animals, insects and birds.

The most unexpected and heartbreaking lesson from the oak trees—one I never expected when I started this book—is a foreshadowing of the future. The ancient oaks once seemed as permanent as the mountains, but their demise is an omen of what will happen to the rest of the environment. In 2002, scientists discovered the majestic Giant Redwoods, which grow no where else on earth, except the northern coastal area of California, also have the oak sudden death fungus.

In the spring of 2003, the sudden death fungus struck seven hundred miles away. In Oregon, southeast of the city of Portland a large nursery found the disease on its rhododendrons and viburnum. Not only can oaks host the disease, but it can also damage azaleas, Douglas fir, bigleaf maple, manzanita, and camellia.

We must heed the seriousness of global warming. Nature's laws operate everywhere. They are constant, strong, and unwavering. No place on earth can they be disregarded without consequences. Like the beautiful California oaks that suddenly die, nature rarely warns us of the consequences that are coming. Scientists do that. The consequences arrive as silently as the way nature dispels her wisdom.

CHAPTER 47

THE EMERALD FOREST

Here in the Pacific Northwest the emerald forest near my home beckons. Alongside trees dripping with emerald green moss, giant waist-high ferns carpet the forest floor. Between the ferns a parade of ever changing flowers appear during spring and summer. The wild flowers politely take their turn, blooming every few weeks in succession, giving each flower its moment in the sun. The blossoms share the spotlight with a flower of a contrasting color, an opposite on the color wheel as a reminder of the many subtle ways the earth seeks balance. Purple violets begin the parade beside yellow Johnny-jump-ups. By Easter chartreuse maidenhair ferns grow freely amongst lacy magenta geranium. When the sweet purple lilacs bloom, yellow skunk cabbage, rich in stinky sulfur permeates the banks of the stream. The parade continues with a variety of mushrooms, some too beautiful and too bright a red to possibly be edible.

The forest offers much food for its inhabitants: licorice ferns, dandelion greens, choke cherries, chicory, chamomile, Blackberries, wild rye, Oregon grape, and pinecones. There are fruit trees—wild cherry, plum, and hazelnuts—that provide a feast for birds and animals. By July and August in an opening a foxglove is glimpsed behind a maple tree, blue bachelor-buttons, daisies, and sweet pea blossom before autumn leaves are ablaze with color.

The forest is alive with squirrels, cooper hawks, flickers, owls, woodpeckers, bats, deer, raccoon, mice, moles, and possum, and all are watched over by a pair of bald eagles that have nested near the lake.

Vibrant, year-round green mosses on the sides of trees come in many varieties. Some resemble miniature pine trees, others look like lamb's wool, but all are soft as baby blankets. In the dry season the moss clings to the tree, sucking moisture from the bark, but when it has plenty of rain, the moss peels easily in big sheets.

Some mornings mist visits between the trees then the wind moves the clouds to let sunlight shimmer on the mist, turning moist air into ribbons filled with tiny prisms. In the forest towering maple, alder, and Douglas fir reach toward the sky. Below the upper story, dappled light plays on the medium-sized trees of hawthorn, wild cherry, hazelnut, and holly.

Music fills the air from songbirds that fly from branch to branch. Now and then the birds land on a rock then drink from the water that sounds like a piano being played as the stream cascades down over the boulders.

In late fall when the trees are bare, the wind moves the branches that ascend well over a hundred feet into the air. The bare branches clap against each other making a sound reminiscent of a million castanets. Walking through the forest it appears that some magic force planted the trees in rows, but it's an illusion. What has happened is that seeds have come to rest upon a dead, fallen log. The saplings followed the straight lines of the mother log, nourished by compost from the decaying tree. That is why in the forest some trees stand as if arranged in a tidy row.

There's much variety in this kingdom of the Standing Ones. They are partners with human beings who walk the well-worn trails. Both have a talent for growing tall and standing erect. Together both types of Standing Ones make magic air that allows life to continue. In the various kingdoms of the Standing Ones, the bark of the trees, which is their outer skin, contains five representative colors: white birch, red pine, black ebony, yellow bamboo, and brown maple.

In places where Standing Ones dwell in groves, they often cluster within natural boundaries wherein they share common elements that come from living on the same soil. Through the cold and dark winter nights, as well as in the laughter of summer's warm glow, through thick and thin, they stand together. When it showers, they reach out their limbs like open arms to catch the gift of the Rain Giver. Their roots are their savings, where they store carbon for photosynthesis and moisture against a time when they are left high and dry.

Among the Standing Ones, ancient communities abide alongside younger, newer groves as a reminder that life constantly reinvents itself. The neighborhoods where Standing Ones live vary from wild to pastoral, from jungle to island paradise, from city street to virgin forest. Yet, in the forests, few virgin trees remain.

Within the groves, individual trees of various ages, sizes, shapes, and talents live side by side. Some are tall, friendly giants like the Sequoia. A few are as thin as a reed, as supple as the willow, as grumpy as crabapples, as clinging as wisteria, as dangerous as the hemlock, and as ancient as Sequoia, a cousin of the Bristlecone pine.

The females of the Standing Ones are found in the pink petals of the magnolia, the red on the dogwood, the sweet smell of lilacs, the blossoms on the wild cherry, and the nourishing chestnuts. The males are found in sheltering oaks, the protective redwoods, in the traveling coconuts, and the spicy pepper trees.

Each community shares the sunlight, rain, and air, understanding that these are necessities of life and to tamper with them is a grievous wrong. So it is that larger trees with power behind their mighty weight and lofty stature do not try to overpower the littler trees. No one is lesser or better in the kingdom of the Standing Ones for all respect nature's holy, unwritten law that speaks of harmony and balance. The law commands that each has a gift to give back to the community, and all will benefit, if each living thing in the forest is free to give its gift. So it is that the larger trees, with a need for more natural resources, have a responsibility to ensure younger ones have a fair chance. To that end, all the trees are designed so their bark runs vertically, and in the rain excess water falls between crevices in the bark, rolling to the ground where it flows into an underground pool. From the pool, all the trees share the new soil formed when the leaves from different trees mingle, and form compost to benefit the entire forest.

The Standing Ones can be seen in their strong, quiet stance turning their leaves to listen to the wind. Although they are excellent listeners, the Standing Ones speak their own sound that whispers to the forest dwellers. When they speak, their leaves move as lips of humans move when they talk. When the wind moves the poplars, the poplar's leaves rustle like taffeta and tell of dancer's skirts. The oak leaves echo the sound of parchment pages being turned in an old book. In cathedrals of Douglas fir, winds gather and roll back the sound of time.

Inside the calm, hushed shade of the forest, humans look up to the sky, as sweet breezes mingle with sometimes sad petitions. Every voice is heard for the branches make ladders whereby human petitions ascend to heaven. In the hush, it's possible to listen to the whispering trees. In the silence, it's possible to hear answers to our questions.

Each grove is a choir that sings in garments of emerald green. With their voices raised on high, the trees push the winds around the beautiful blue planet, keeping it cool and safe, so we can enter through silver mists and take our place in the kingdom of the Standing Ones.

CHAPTER 48

THE LOVER'S FOREST

At the beginning of the book, you may recall Nick telephoned to say that he had found his soul mate. Then while I was on a walk, noisy crows drew me to a cork oak that was surrounded by ancient California live oaks. A few months later, Nick got engaged, but before the wedding, the couple broke their engagement. Shortly thereafter, work took Nick away from California, and last I heard she married a super guy.

Thinking of engagements reminded me of my own betrothal. For an engagement ring, I favored a simple apple-green jade ring that had belonged to my fiancé's mother. Sentimentality surrounded the ring since his mother had died when he was young. My heart seemed to open like a rose when I first ran my fingertips over the oak tree carved into the green jade. The ring appealed to something deep within me long before I ever thought of writing this book. Loving color as I do, I preferred the green jade to a traditional diamond.

I didn't realize it then, but the gems that adorn rings have an interesting connection to the carob tree. The Greeks recognized that, unlike olives whose pits can vary in size, the inedible part of the carob seed always had the exact same weight, 0.2 grams. Weighing objects as precious and as tiny as fragments of gemstones demands a precise, very small, uniform unit of weight. To meet the need for accuracy, early gem traders turned to carob seeds, from which the word carat evolved. A one-carat diamond weighs 0.2 grams, the same as a carob seed.

Down through time, couples have treasured gems given in affection. Now and then I trace the outline of the oak that an unknown artist patiently carved into the hard green stone of my engagement ring. Then my imagination can transport me to an image of a couple standing arms entwined under the shade of an oak tree. The tree in my mind's eye is real, having started hundreds of years ago in an incredibly romantic spot. It's reached by way of a narrow footpath at the base of the Santa Monica Mountains. At the end of the trail that follows a shallow streambed, there's a sacred Chumash Indian site among the lava outcroppings. A profusion of ancient coastal live oaks grow there. Typical of the wild oaks left to nature's artistry, the branches drape down until they touch the earth. For centuries the oaks dropped their leaves which eventually formed deep, rich loam. As a result the ground under the trees is soft and sweet, especially after spring grasses spread a green blanket on nature's bed. It's possible to pull back the branches, as if pulling back a curtain, and walk into the area surrounding the tree trunk. It is like entering a large secret room, a place of retreat into a private natural world where the sun and moon peek at lovers through the oak leaves. Bushtits sing their hushed, yet melodious song to the lovers.

A magical connection exists between trees and love. The branches of a tree reach out like someone offering a warm hug. From the sprouting anew of blossoms in spring, trees represent the ability of love for renewal. A tree's immensity and longevity represents the power of love to endure.

Mother Theresa said that loving and being loved is our sacred purpose here on earth.

It would be thrilling and also very healing to the environment, if around the world, lovers planted a tree together as a symbol of their love. If the relationship did not work out, it wouldn't matter; the tree would remain standing as a tribute to the awesome ability within each of us to love and be loved.

Valley oak, Malibu Canyon, CA

A variety of other reasons present themselves for planting a tree as a symbol of love. One can plant a tree to honor one's parents, a beloved friend, siblings, and the family pet. The planting might coincide with celebrating a wedding day or the birth of a child. If commemorating someone deceased, long after a beloved has left this earth the tree would live on. Such a tree would become a living reminder that love endures and transcends death. In a similar way, millions of years ago, dead trees transcended death by turning into coal. For years, coal has warmed homes as love has warmed the human heart.

The inherent benefits trees give are powerful and enduring, reminding us of the awesome possibilities we also hold within our being. If trees planted to celebrate love were given a recognizable heart emblem, it would serve as a symbol of our ability to create love and beauty in this world. Such an emblem would be one that didn't harm the tree. It might be a plaque at the tree's base, or a heart painted on the tree's trunk. For love's souvenir, people could also plant one of the several varieties of trees with heart-shaped leaves. For instance, the Indian flowering bean tree called the Catalpa is one option. From the Himalayas has come the Morus Nigra, better known as the Black Mulberry. There are other trees with heart-shaped leaves, such as the Linden or Basswood, which was an ancient symbol of fidelity. Its flowers provide a fragrance as haunting as an unforgettable love. North America offers lovers the Eastern Redbud with hearts for leaves and flowers as red as a bouquet of roses. The beautiful Princess Tree or Royal Paulowna blooms with long purple, trumpet-shaped flowers that resemble foxglove, the plant used to strengthen the heart. Another option is to plant lilacs, the Victorian symbol of love's sweet beginnings.

Hopefully, people would think twice before cutting down any tree that represented our ability to love and be loved. Over time, the planting of these trees of various species, shapes, and sizes would be scattered throughout the continents. They would grow in cities, on farms, on hills, and in valleys. By planting such trees present and future generations would create for the earth a lasting, living forest of love.

Selected Bibliography

McLuhan, T.C., Ed., *Touch the Earth: A Self-Portrait of Indian Existence*, (New York: A Touchstone Book, Simon & Schuster, Inc., 1971, 47).

Edwards, Tyron, D.D., *The New Dictionary of Thoughts, A Cyclopedia of Quotations*, "Nature, Goethe," (New York: Standard Book Company, 1955, 419).

6. SOUL MATES

Zimmerman, J.E., *Dictionary of Classical Mythology*, (New York: Bantam Books, 1964, 40-41).

10. THE SKATER'S ATTITUDE

Kenworthy, MaryLou, "Thought Matters," http://www.thoughtmatters.net/, (February 1, 2004)

11. LEAVES of RED and GREEN

Evans, Patricia, *The Verbally Abusive Relationship: How to Recognize It and How to Respond*, (Avon, Massachusetts: Adams Media Corp., 2nd ed., March 1996).

Kilmer, Joyce, *Trees*, http://www.cs.alfred.edu/~lansdoct/poems/trees.html (3/16/2001).

12. KARMIC RELATIONSHIPS

Brockman, Frank C., *Trees of North America: A Guide to Field Identification*, (New York: St. Martin's Press, 2001).

Brunner, E, "Sullivan Tree Identification Key," *Sullivan Middle School*, Sullivan, Illinois, *http://home.sullivan.k12.il.us/ teachers/brunner/tree/* (July 26, 2003).

17. NATURE'S ART

Bach, Richard, *Jonathan Livingston Seagull*, (New York: Avon Books, 1973).

Breslin, James A., *Mark Rothko: A Biography*, (Chicago: University of Chicago Press, 3rd Ed., 1998).

Edwards, Tyron, D.D, Ibid, "Nature, John Newton," p.418.

____"Spiders Spider Silk," *Journey's to Wild Places Docent*, *http://www.szgdocent.org/ff/f-ssilk.htm* (February 12, 2004).

Henahan, Sean, "Bioengineered Spider Silk," *Access Excellence*, *http://www.accessexcellence.org/WN/SU/spider.html* (March 3, 2001).

Ottman, Klaus, *The Essential Mark Rothko*, (New York: Harry N. Abrams, 2003).

Trivedi, Bijal P., "Lab Spins Artificial Spider Silk, Paving Way to New Materials," *National Geographic News*, *http://news.nationalgeographic.com/news/2002/01/0117_020117TVspidermammals.html* (June 14, 2003).

Tierra, Leslie, *The Herbs of Life: Health and Healing Using Western and Chinese Techniques, Dandelion*, (Freedom, CA: The Crossing Press, 1992).

____"Great Quotations", (Anne Frank), *http://www.cybernation.com/victory/quotations/authors/quotes_frank_anne.html* (July 9, 2003).

18. THE AMAZING SEED

Baradat, Leon P., *Political Ideologies*, 2nd Edition, (Englewood Cliffs, New Jersey: Prentice-Hall, Inc., 1984).

Barzun, Jacques, *From Dawn to Decadence: 500 Years of Western Cultural Life*, (New York: Harper Collins, 2000).

Brians, Paul, "The Enlightenment," *Washington State University*, *http://www.wsu.edu:8080/~brians/hum_303/enlightenment.html* (July 18, 2003).

Commager, Henry Steele Commager, editor, *Documents of American History*, 7th Edition, (New York: Appleton-Century-Crofts, A Division of Meredith Publishing Company, 1963).

Encyclopedia Britannica, 15th ed., Macropaedia, s.v. "Darwin," Chicago, 1979.

____*Ibid.*, Micropedia, s.v. "Divine Right of Kings."

Gillham, Nicholas Wright, *The Life of Sir Francis Galton: From African Exploration to the Birth of Eugenics*, (Oxford, England: Oxford University Press, 2001).

Hart, Albert Bushnell, *American History Told by Contemporaries: Era of Colonization*, Vol.1, (New York: Machmillan Company, 1916).

Heschel, Abraham Joshua, *The Insecurity of Freedom: Essays On Human Existence*, (New York: Shocken Books, 1966).

Tredoux, Gavan, Editor, "Francis Galton," *Galton.org*. http://www.mugu.com/galton/start.html (March 3, 2002).

Watters, H. Barbara, *Sex and the Outer Planets*, (Washington, D.C.: Valhalla Paperbacks, LTD., 1971, 28-38).

19. CREATIVITY

Lieberman, Max, McFadden, Michael, "Divine Inspiration in Art," http://www.sol.com.au/kor/9_01.htm (June 20, 2003).

23. SING a HAPPY SONG

MacIvor, Virginia and LaForest, Sandra, *Vibrations: Healing Through Color, Homoeopathy and Radionics*, (New York: Samuel Weiser, Inc, 1979).

Robertson, Don, "About Positive Music: The Plant Experiments," http://dovesong.com/postive_music/plant_experiment.asp (May 24, 2003).

"French Physicist Creates New Melodies—Plant Songs," http://wwww.earthpulse.com/science/songs.html (May 24, 2003).

29. LEAVING SOMETHING BEHIND

Lombardi, Frances G. & Gerald Scott, *Circle without End*, (Happy Camp, CA: Naturegraph Publishers, 1982, 38).

30. GIVING BACK
Keyes, Bruce, "History of Brigham City Cemetery," http://users.evl.net/~hmltn/cemeteries/brigham_city_cemetery_history.htm (September 20, 2003).

33. THE ANCIENT OAK
Lee, Cynthia, "Petition's Validity Is Questioned," *The Thousand Oaks Daily News,* January 8, 1990, p.1, 2.

Mansfield, John, Exec. Producer, "Preserving the Balance: Tropical Forest," *Nova Television: Adventures in Science, WGBH Boston* (Boston: Addison-Wesley Publishing Company, 1983, 192-3).

Maravilla, Nach M, *Daniel Ludwig: The Invisible Billionaire,* http://www.powerhomebiz.com/Success/ludwig.htm, (May 10, 2003). Originally published: Shields, Jerry and Shields, Lerry, *The Invisible Billionaire: Daniel Ludwig,* (Boston: Houghton-Mifflin, 1986).

Watkins, Thayer, "The Jari Project," *San Jose State University Economics Department,* http://www2.sjsu.edu/faculty/wakins/jari.htm (February 20, 2002).

34. THE SEED REPRODUCES ITSELF
Spink, Kathryn, *Mother Teresa: A Complete Authorized Biography,* (San Francisco: HarperSanFrancisco, 1998).

____Bruce, Patsy, "Willie Nelson," *American Profile, Celebrating Hometown Life,* http://www.americanprofile.com/issues/20030831/20030831_3300.asp. (February 2, 2004).

36. AN EARTHLY WOMAN
Brown, Kathleen M., "Jamestown Interpretive Essays: Women in Early Jamestown," http://www.iath.virginia.edu/vcdh/jamestown/essays/brown_essay.html (February 12, 2004).

____"Indentured Servants and Transported Convicts," http://www.stratfordhall.org/ed-servants.html EDUCATION (January 6, 2004).

Amon, Aline, comp, *The Earth Is Sore: Native Americans on Nature*, (New York: Atheneum, 1981).

____"England and Irish Slaves," *http://www.ewtn.com/library/ HUMANITY/SLAVES.TXT* (October 23, 2003).

Crane, Elaine Forman, *Ebb Tide in New England: Women, Seaports, and Social Change, 1630-1800*, (Boston: Northeastern University Press, 1998).

Hensel, Tabatha, "Going Native: Inter-Cultural Relations Between Europeans and Indians." *http://muweb. millersville.edu/~papers/hensel.html* (January 29, 2004).

____"Women's Property Act, 1852," *Special Collection University Archives,* Rutgers University Library, *http:// womenshistory.about.com/library/ency/blwh_property.h*tm (March 2, 2001).

Sams, Jamie, *"Earth Medicine: Ancestors' Way of Harmony for Many Moons,* (San Francisco: HarperSanFrancisco, 1994).

37. WITHOUT ONE, NO OTHER

____"No Man is an Island," *Essaybank.CO, http://www.e ssaybank.co.uk/free_coursework/1237.html* (May 12, 2003).

Shakespeare, William, *A Midsummer Night's Dream*, (New York: A Signet Book, The New American Library, 1964).

____"Rainforest Curriculum, Fact Sheet, Wonders, and Destruction," *http://www.earthsbirthday.org/rainforest_curr/* (January 19, 2001).

Hui, Stephen, "Deforestation: Humankind and Global Ecological Crisis," *http://www.Sfu.ca/~shui/resources/ deforestaion.htm* (September 28, 2000).

____"The Destruction of Oceanic Phytomass," *http:// www.geocities.com/carbonomics/Mcsppub/11sp06/ 11sp06j_f.html* (February 11, 2004).

Walsch, Neale Donald, *Conversations with God, an uncommon dialogue,* Book 3, 291 (Charlottesville, Virginia: Hampton Roads Publishing Company, Inc., 1998).

38. EGGS IN THE NEST

Frank, Jeff, "OM . . . OM . . . OM," *http://www.greenguerrilla.com/om.htm* (March 24, 2001).

Lefevre, Greg, San Francisco CNN Bureau Chief, "Despite Efforts, Ozone Layer Will Take a Long Time to Heal," *CNN.com, NATURE, http://europe.cnn.com/2000/NATURE/12/16/oxone/index.html* (December 18, 2000).

Lemonick, Michael D., "Answers to Global Warming Are in the Wind, How to Prevent a Meltdown," *TIME*, Special Edition, Earth Day 2000, April-May, 2000, 61-63.

____"Life Under the Hole in the Sky," *Salon.com, Health, http://www.salon.com/health/feature/2000/11/03/oxone/index1.html* (March 17, 2001).

MacKeen, Dawn, "Global Warning," *Salon.com, News, http://www.salon.com/news/feature/2001/03/02/warming/index.html* (May 3, 2001).

39. THE FOREST OF THE FUTURE

Curtis, Bob, "Sierran Montane Forests," *eNature.com, http://ww.enature.com/habitats/show_sublifezone.asp?sublifezoneID=42* (September 30, 2003).

Hymon, Steve, "Not So Clear-Cut a Choice," *Outdoor Explorer, http://www.outdoorexplorer.com/hike/clearcut.html* (February 22, 2001).

Landis, Scott, "Seventh Generation Forestry: Menominee Indians Set the Standard for Sustainable Forest Management." *http://www.menominee.edu/sdi/Seventh GenerationForesty.h*tm (March 10, 2001).

Donnelly, Michael, "Amazing Facts," *University of Oregon.edu, http://darkwing.uoregon.edu/~recycle/Factoids_amazing_facts.htm* (February 26, 2004).

40. A GIFT TO GIVE

____"Chumash Diet, Food, and Medicine," *Santa Barbara Natural History Museum, www.sbnatur.org/research/anthro/chumash/health.htm* (March 21, 2003).

Lind, James, "A Treatise of the Scurvy, 1753," *http://WWW?Nautica/Medicine/Lind(1753).html* (May 22, 2003)

Weatherford, Jack, *Indian Givers: How the Indians of the Americas Transformed the World*, (New York: Faucett-Columbine, 1988, 177-196).

Plotkin, Mark, "Nature's Gifts: The Hidden Medicine Chest," *Time Magazine*, Special Edition, Earth Day 2000, April-May 2000, 34.

____ "Frequently Asked Questions," *Stem Cell Research Foundation http://www.cordblooddonor.org/* (February 18, 2004).

41. THE TOXIN GOBBLERS

Wolverton, Bill, *How to Grow Fresh Air: 50 Houseplants That Purify Home or Office,"* (New York: Penguin, 1997).

Wolverton, Bill, "Sources of Chemical Emissions," *http://www.wolvertonenvironmental.com/chem.htm* (July, 2000).

____ "Facts: Plants are Nature's Most Efficient Environmental Air Cleaner," *http://www. plants4cleanair. org/-facts.html* (March, 2000).

____ "How to Breathe Easier at Home and in the Office," *Clean Air Council, http://www. accentplantscaping. com/ CleanAirCouncil.html* (May 23, 2003).

"Clean the Air with House Plants," *http://ablewoman.org/ 9703/plants.htm* (May 23, 2003).

42. SCALES OF BALANCE

____ "Chinese Cultural Studies: Yin and Yang in Medical Theory," *http://acc6.its.brooklyn.cuny.edu/~phalsall/texts/ yinyang*, (May, 2003)

Hay, Louise L., *Heal Your Body: The Mental Causes for Physical Illness and the Metaphysical Way to Overcome Them, "Spinal Misalignments,"* (Carson, California: Hay House, 3rd ed., 1984, 76-80).

____ "Figures of Justice," *Supreme Court.gov, http:// www.supremecourtus.gov/about/figure . . .* , (Feb 26, 2004).

____ "Plankton: Life Giving Miracle of the Ocean," *Planet*

Ocean Journal, http://www.plantoceansociety.com/journal-plankton.html (May, 2003).

———"Checkup for the Planet: Winds of Change, What's the Connection between Hurricanes and Global Warming?" *Greenpeace Magazine*, Winter, 1999, 7-12.

Graffis, Mark, "Arctic Climate Changing Rapidly," *Environmental News Service*, http://www.hartford-hwp.com/archives/23/005.html, May, 2003.

Mathews-Amos, Amy, Bernstson, Ewann A., "Climate Change Harms Ocean Life," *http://pnews.org/art/3art/climate.html*, May, 2003

——— "Going, Going, Gone! Climate Change and Global Glacier Decline, *World Wildlife Fund*, www. worldwildlife. org/?climate/climate.cfm?sectionid=111&newspaperid=16?-?18k?-?Feb25,2004, (February 12, 2004).

44. NATURE SHOWS THE WAY

Golembeski, Dean J., "Trees Mother Nature's Diaries," *Troika Magazine*, *http://www.troikamagazine.com/fiction/bodyoffacts_1994.html* (March 24, 2001).

———"Bark Beetles," *University of California Pest Management Guidelines*, http://www.ipm.ucdavis.edu/PMG/PESTNOTES/pn7421.html, (September 13, 2003)

——— "Ringers," *Whyfiles*, http://whyfiles.org/o21climate/ringers.html (November 24, 2000)

Keslick, John Jr., "Coarse Woody Debris-Water/Moisture," *Tree Biological Laboratory, Allegheny Defense Project*, http://www.chesco.com/~treeman/sound/soundscience/water.html, (May 3, 2003).

———"Shades of Green: Earth's Forests, Temperate," *Thinkquest.org*, *http://library.thinkquest.org/17456/temperate2.html*, (January 23, 2001).

45. THE DROUGHT

———"Return to Genesis of Eden? The El Niño Gallery: A

Documentation of Disasters," *http://dhushara.tripod.com/ book/diversit/extra/nino.htm* (January 17, 2001)

____"El Niño's Surprisingly Steady Pacific Rains Can Affect World Weather," NASA/Goddard Space Flight Center, *http://www.sciencedaily.com/releases/2003/08/030821072404.htm* (February 12, 2004)

Krautkraemer, John and Willey, Zach Dr., "Learning From California's Drought, *Environmental Defense Council*, adapted from an article in the San Diego Union. *http://www.edf.org/pubs/EDF-Letter/1991/Jun/j_drought.html*, (January 10, 2001)

____"La Niña Summary," *National Weather Service Forecast Office, http://www.crh.noaa.gov/mqt/index.php?page=climate/lanina/index,* (February 29, 2004).

____"About ARM, Global Warming," *The Atmospheric Radiation Measurement Program of the U.S. Department of Energy, http://www.arm.gov/docs/education/aboutarm.html* (February 29, 2004).

____"Natural History of Mono Lake," *Mono Lake Committee, http://www.monolake.org/naturalhistory/index.html,* (January 5, 2001)

46. ALL THINGS CONNECT

____"Sudden Oak Death," *Marin County University of California Cooperative Extension, http://cemarin.ucdavis.edu/index2.html,* (May 10, 2001)

____California Department of Water Resources Drought Preparedness, *http://watersupplyconditions.water.ca.gov/next_drought.cfm* (January 8, 2004).

____La Niña Summary, *National Weather Service, http://www.crh.noaa.gov/mqt/index.php?page=climate/lanina/index,* (February 11, 2004).

____"Children of the Tropics: El Niño and La Niña," *UCAR Communications and Facts, http://www.ucar.edu/communications/factsheets/elnino/,* (February 11, 2004).

———"Global Warming," *Atmospheric Radiation Measurement Program, U.S. Department of Energy,* http://www.arm.gov/docs/education/globwarm/causglobwarm.html (February 20, 2004)

Ackerman, Tom, Chief Scientist U.S. Department of Energy, "Why Should We Care About a Few PPM of CO2?", *Department of Atmospheric Radiation Measurement Program,* http://www.arm.gov/docs/education/globwarm/ch_scientist/sld033.htm, (February 20, 2004).

———"Tropical Rainforest Facts," National Resources Defense Council, *http://www.nrdc.org/land/forests/frainf/asp,* (July 9, 2000).

BVG